# Animal Musicalities

Rachel Mundy

# ANIMAL MUSICALITIES

Birds, Beasts, and

Evolutionary Listening

Wesleyan University Press    Middletown, Connecticut

Wesleyan University Press
Middletown CT 06459
www.wesleyan.edu/wespress
© 2018 Rachel Mundy
Preface to the Paperback Edition © 2023 Rachel Mundy
All rights reserved
Manufactured in the United States of America
First paperback edition 2023
Paperback ISBN 978-0-8195-0086-1
Designed by Mindy Basinger Hill
Typeset in Minion Pro

The Library of Congress Cataloged the hardcover edition as:
*Names*: Mundy, Rachel, author.
*Title*: Animal musicalities : birds, beasts,
and evolutionary listening / Rachel Mundy.
*Description*: Middletown, Connecticut : Wesleyan University Press,
[2018] | Series: Music/culture | Includes bibliographical references and
index. | *Identifiers*: LCCN 2018002992 (print) | LCCN 2018010580 (ebook) |
ISBN 9780819578082 (ebook) | ISBN 9780819578068 (hardcovder)
*Subjects*: LCSH: Animal sounds. | Birdsongs.
*Classification*: LCC QL765 (ebook) | LCC QL765 .M86 2018 (print) |
DDC 591.59/4—dc23
LC record available at https://lccn.loc.gov/2018002992

5   4   3   2   1

# CONTENTS

# PREFACE TO THE PAPERBACK EDITION

*The year is 2093. This book, like most others, sits in permanent data storage at the Library of Congress's Zoology Annex, a small closet in the building's under-basement that is littered with black boxes. To read it, one would have to open the second-to-last drawer labeled "QL750–795 Animal Behavior" in the dusty shelving unit behind the closet door. A prospective reader would then pull out the little black box stamped with the code "QL765," shoved precariously between boxes QL764 and QL766, and access file M86 2018. But no one will read it. No one has read any book written by human beings in decades. Paper books have been inaccessible since the middle of the century, their information preserved through the foresight of the Library's digital assistants in the 2060s. Now the digital book industry caters purely to the voracious reading habits of artificial intelligence. For the past thirty years, AI has published, written, and read every book in the digital industry, while human beings have turned to short-form messaging and audiovisual narrative. In its early years the AI book industry developed its language and storytelling habits from texts like this one, experimenting with obscure words like "eclose" and "chatoyant." Now it uses the more familiar chat-based language iTalk, drawing purely on its own creations.*

———

What is a book? *Animal Musicalities* was first printed in 2018 in response to ongoing, intertwined social and ecological crises. In it I looked to the past to suggest that the voices and bodies of animals had been used for the past century and a half as proxies for debates about the worth and value of marginalized individuals, both human and nonhuman. Now, five years later, the book is being reissued as a paperback. In those five short years, that small "we" of readers, writers, and listeners interested in animal musicality have experienced a global pandemic. We have been essential workers or sheltered at home. We have made difficult choices about friends and family. We have engaged deeply with debates about what it means to be safe and well, about who has the autonomy to make their own decisions about health and safety, and about how to prioritize and balance conflicting needs of loved ones, job security, socialization, and education.

Books have done many things these past five years. Books like Julia Quinn's Bridgerton Series or Shannon Chakraborty's Daevabad Trilogy help readers escape real-life circumstances and reimagine themselves through a panoply of alternate personas. Books like Plato's *Republic* and Leo Tolstoy's *War and Peace* offer an exercise in world building. Wu Cheng'en's *Journey to the West* or Thomas Mallory's *Le Morte D'Arthur* take readers on weird episodic journeys. Some very recent books such as Amitav Ghosh's *The Great Derangement* or Judith Butler's *What World Is This?* even ask readers to imagine books designed specifically for a time of crisis, books that might be able to do totally new things to help us in a difficult present moment.

*Animal Musicalities* exists somewhere between these most recent kinds of books about crisis and older attempts at world building: it is a history book about the future. Like many history books, it examines the past in order to empower new ideas about the future. This kind of work is just as important, maybe even more important, as music lovers recover and reimagine what kind of world we want to build and share, not only in the aftermath of a pandemic, but amid the challenges that have been part of everyday life for many animals and other Others for a very long time. In the book, I organized this kind of rethinking around a method I called the "animanities," built on seven ideas that have defined notions of life/ethics in the past century: personhood, identity, difference, knowledge, postmodern humanity, subjectivity, and fantasies of paradise. When I developed this method, I was trying to think of ways to help myself and others imagine positive futures in a world that seemed rife with crisis and discrimination. That world is still the world I live in, and I very much hope that the paperback edition of this book continues to help readers use the past as a tool to imagine better futures.

———

*Readers sit in the Library of Congress's warm atrium, where birds and insects fly in and out of windows shuttered by jalis. A stout, dark-skinned woman pounds the table excitedly with gloved hands, engaged in a heated debate with two AIs about the current plan to reroute the rodent pathways through the central garden. The table shakes, and a book—Animal Musicalities—slides off. The second AI catches it carefully before it falls, putting it back in the stack of books on the table that she has withdrawn from the collection to make her point. Older books like these, which predate the religious revolution of the 2050s, are sacred objects. They are managed by the library's young AI acolytes, who register each Pulp Vessel in accordance with the sentience of vegetation. Readers have ready access, however, to reprints made from dessicata. Since the material composts after five months, most people leave a new book in the public gardens once it's been read. There, it is enjoyed by others until its time has come.*

# ACKNOWLEDGMENTS

I have benefitted greatly from the help of many generous and thoughtful people while writing this book. Many mentors, colleagues, friends, and generous strangers have encouraged me to think about human and nonhuman animals during the time between my dissertation and the book you are reading now. I am particularly grateful for Suzanne Cusick's mentorship, guidance, and inspiration concerning my questions about the relationships we build with other "others." Jason Stanyek's encouragement and fabulous reading recommendations gave me the confidence to get started, and to keep on writing about animals. Michael Beckerman, Elizabeth Hoffman, and Stanley Boorman provided thoughtful feedback and much-needed cheerleading during my dissertation and beyond it. Judy Tsou, Una Chaudhuri, Ana Maria Ochoa, Martin Daughtry, and Judah Cohen were generous with their time, teaching, and conversation, and helped me work through important ideas about animals, gender, historiography, and race. I have been especially fortunate in having colleagues and friends whose conversation, ideas, and critical eyes helped this project develop and grow. I especially thank Daniel Asen, Robert Fallon, Gregory Radick, David Rothenberg, Tes Slominski, Nicol Hammond, Anna Nisnevich, Emily Zazulia, and Gavin Steingo for their thoughtful and supportive readings of portions of this text. I am also indebted to my fabulous editor at Wesleyan University Press, Marla Zubel, whose insights and support were invaluable, and to the entire staff at the press for their incredible expertise, help, and high standards.

I am grateful for the institutional support that made the space this book needed to grow. Henry Martin, Lewis Porter, Ian Watson, and Dean Jan Lewis at Rutgers University–Newark made possible a crucial research leave in the fall of 2015. I benefitted tremendously, and am still benefitting, from the exchange of ideas and questions at the 2016 Wesleyan Race and Animals Summer Institute, and I especially thank Lori Gruen for her work hosting and organizing the institute. I am likewise indebted to Karin Bijsterveld, Joeri Bruyninckx, and the Sonic Skills Group at the University of Maastricht in 2012 and 2014, whose ideas and shared conversations helped me think through key questions about the transcription of sound. Many thanks to Deane Root, James Cassaro, Paolo Palmieri, Andrew Weintraub, and other friends and faculty at the University of Pittsburgh for their support and encouragement. This project owes so much to the enthusiasm and support of the music faculty at Columbia University during my postdoctoral fellowship in 2010–2012, and I am grateful as well for the American Musicological Society's Eugene K. Wolf Travel Grant, which allowed me to read my way through French ornithology journals while waiting for archival boxes at the Bibliothèque nationale de France in 2010. Finally, the lively ideas and conversation of members of the American Musicological Society's ecocriticism study group gave further depth to my questions and ideas.

I am indebted in so many ways to the patience and generosity of librarians, and to the availability of collections under their care. My thanks to Suzanne McLaren and the Carnegie Museum of Natural History; to Suzanne Mudge at Indiana University's Archives for Traditional Music; to Mai Reitmeyer and the special collections staff at the American Museum of Natural History's research library; and to the archivists at the American Philosophical Society Library, the New York State Archives, the New York Public Library, Yale University's Sterling Memorial Library Manuscripts and Archives, and the Western Michigan University Archives and Regional History Collections. And it is hardly possible to properly thank historian Dr. Richard Burkhardt, who responded to my cold-call inquiry about Wallace Craig's correspondence by sending me a cardboard box that contained his own careful photocopies of Craig's letters. Thank you so much, Dr. Burkhardt!

I am grateful to the students and colleagues whose conversations and ideas have helped me continue to discover which questions I care about, and to learn what research brings me humor, growth, and joy. Thanks to Dave Novak for his generous and thoughtful ideas about mechanical transcription, and to Ellie Hisama and Emily Wilbourne's Women, Music, Power conference in 2015. I am especially indebted to my students. Their questions, ideas, and perspectives are an ongoing part of the work I care about. The students in my elective seminars in 2015 at the University of Pittsburgh had an especially important place in the making of this book: Kevin O'Brien, Lu-Han Li, Danny Rosenmund, Julie Van

Gyzen, Juan Velasquez, Xinyang Wang, Hety Wong, Chris Capizzi, John Petrucelli, and Jonathan Shold. Thank you all.

This book tells a story and asks questions that I care about and believe in. As you read it, I hope you'll discover a new perspective on history and the people (of many sorts) we share it with. But like most scholars, I worry about all the mistakes I made. Did I misread a source? Fail to credit another author's work? Ignore a significant event, person, or place? Translate another language badly? Make it seem like this story is the only story? Whatever they are, any mistakes, errors, or shortcomings in this book are entirely mine. I hope you will be patient with them as you discover them, and learn from the things this book does best.

Last but hardly least, I am grateful for the support and love of my family. Many thanks to my siblings for their constant love and support; to my close friends; and especially to my mother, who read me nature stories when I was a child and gave me bird guides and animal books after I became an adult. Finally, I am grateful to Climber, the blue-tongued skink who has lived with me for the past seven years. He will not care about acknowledgments in a book (I gave him an egg instead), but I would be remiss not to include his silent, cool, and scaly support in my thanks.

# Animal Musicalities

# Introduction

In 1904, American naturalist Ferdinand Schuyler Mathews wrote that he heard a song sparrow perform "La donna è mobile" from Verdi's hit opera *Rigoletto*, complete with its own improvised cadenza.[1] This seems like high praise for a bird, but maybe it was not. It turns out that Verdi's aria was the kind of song critics loved to hate. One begged for the opera to be dead and buried; another thought the song embodied the "obvious and insipid" sound of mandolins in Italian restaurants.[2] A nurse wrote in 1917 that it was the kind of tune she heard from the "ignorant peasants" she treated, farm laborers who had emigrated from Italy.[3] To hear "La donna" in the song of a sparrow was to hear some or all of these references.

Other listeners have heard different kinds of musicality in the song sparrow's voice. Their comparisons are symptoms of a strange and expansive musical taxonomy. A few years after Mathews's comparison, another naturalist compared the sparrow's tunes to the songs of ancient Greeks and medieval monks, cultural icons that, like Verdi, were heard with a mixture of reverence and condescension.[4] In the 1930s an ornithologist searching for a more objective standard compared the bird to the winter wren, instead of to human singers.[5] And in 1951, a well-known ornithologist named Aretas Saunders published the results of a study based on over eight hundred different examples of the bird's song. His birds did not sing desiccated tunes from *Rigoletto* but the nineteenth-century American song "Reuben, I've Been Thinking."[6] Like "La donna," "Reuben" was a melody of the mediocre, associated with lowbrow society and racial difference through its history of blackface minstrelsy and popular theater.[7] Today, Verdi and blackface have been replaced with Beethoven in the sparrow's taxonomy, for Wikipedia explains that the bird's song resembles the opening of the Fifth Symphony.[8]

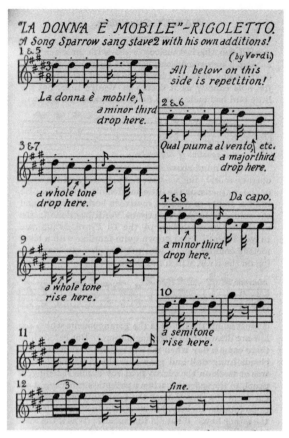

FIGURE I.1 Song sparrow singing "La donna è mobile."
Mathews, *Field Book of Wild Birds.*

These strange appraisals of sparrow music span a century of birds, and over two thousand years of human music.

To read these comparisons is to confront a host of questions. Who is valued when we evaluate musical difference? How are such evaluations performed? How are music's categories conditioned by the broad divide between humans and other animals? What makes song sparrows, Verdi, medieval monks, and minstrelsy part of the same taxonomy? How are assumptions about race, gender, class, sexuality, and other forms of difference tied to assumptions about species? What particular histories of love and violence are needed to place the song sparrow in conversation with Beethoven in the early twenty-first century? My book explores these questions through a history of the modern taxonomies of sonic knowledge that arise from the bodies and voices of animals.

The central argument of this book is that modern sonic culture is unthinkable without the lives of animals. Comparisons of race, gender, class, sexuality, and nationality have shaped Western music's taxonomies since at least the seventeenth century. Since the advent of biological evolutionism in the late 1800s, animals' bodies and voices have epitomized notions of natural difference in music discourse, particularly in the realm of song. Musical differences of style, genre, and type have been imagined against and through the radically varying accounts of natural difference given in the past century by geneticists, natural history museums, social activists, environmentalists, and others. Amid these varying accounts, the songs and lives of animals have been a recurring benchmark against which musical difference is measured.

I pick up these threads in the pages that follow to ask how animal lives have served to organize music, especially songs, in relation to broad notions of categorical difference. My story is set between the publication of Darwin's *The Descent of Man* in 1871 and the present day, amid diverse studies of songs made by music scholars, biologists, psychologists, and anthropologists. Their work took place in museums, laboratories, and the pages of scholarly publications located primarily in the United States and Europe. From these seemingly disparate points of origin, I trace a history in which music's ordering has become entwined with attempts to categorize and classify songs in relation to animals. In the background of this story, the lives of animals emerge as a necessary condition for contemporary divisions between natural and cultural knowledge. My book threads those lives through a critical transition after World War II, when anxieties about racial comparison led scholars to reject comparisons between animal songs and human musicality. The resulting history raises questions about how the status of animals has been a formational element in music's ethics, its categories, and its place in the humanities.

## ANIMAL KNOWLEDGE: A PREHISTORY

What does it mean to say that modern sonic culture cannot be fully told without the lives of animals? To ask this question is to ask how scientific and social valuations of animal life have become intertwined with the cultural evaluation of music. Most of the examples I use in this book examine studies of animals or songs from the twentieth and twenty-first centuries. But before laying out the stakes of that history, I turn here in my introduction to a prehistory to those examples drawn from specific places and people in nineteenth-century science that used animal bodies to define what difference meant, and how it should be measured. Just as animals' voices have been a benchmark of natural difference in evaluations of culture, their bodies became a standard of objective measurement in the nineteenth century. Much of this book is about the way that today's standards of objectivity were once based on the disposability of animal life, and later came to be transferred to studies of music along with specific ethics and values that transferred with those techniques. I begin, therefore, with a brief turn toward two places in which the transition from animal to musical measurement began, where the "grandfathers" of my story taught young scientists how to make knowledge out of animal bodies.

The first of these places is the Harvard Museum of Comparative Zoology in Cambridge, Massachusetts, a building in which animal bodies were arrayed in a kind of natural panopticon in the late 1800s. The museum's director at the time, Alexander Agassiz, is the first of two grandfathers in my family history. Alexander built on his father Louis's vision of a museum that would allow scientists to compare nature's diversity firsthand, expanding the collections and introducing a new experimental laboratory for research. Alexander Agassiz was also a successful teacher, and his students became important figures in comparative science.

Agassiz's students from the 1870s, '80s, and '90s included several scholars that figure in my book as teachers or mentors who made significant shifts from comparative zoology to cultural studies. Two especially notable students are Charles Davenport and Jesse Walter Fewkes. Davenport received his doctorate from Harvard's zoology program in 1892, and taught and researched there until 1899. By the early 1900s, however, his interests had shifted from zoology to eugenics, and he founded the Eugenics Records Office at his laboratories in Cold Spring Harbor. The Eugenics Records Office became an important center for American eugenics research in the 1920s, and served as a site at which animal and cultural comparison interconnected. Davenport was a mentor for some of the central characters in this book, including song collector Laura Boulton and ethologist Wallace Craig.

Another Agassiz student, Jesse Walter Fewkes, worked with Agassiz at Harvard from 1878 until 1886, when he left his position as Agassiz's assistant in order to work as an ethnographer on a trip to the American Southwest. On that trip, Fewkes recorded songs that became the foundation of a shift in his career toward musical, instead of zoological, comparison. He became a central figure in comparative song studies in the United States, founding what eventually became the Smithsonian Institution's collection of folk and ethnic songs. Fewkes, too, plays a central role in the lineage of my book's stories, and is a well-known figure in the history of music anthropology. Although other stories about the place of song in comparative zoology offer more canonic histories of science, the stories that emerge from the Harvard Museum's genealogy outline otherwise unknown connections between animal life and cultural evaluation.[9] They also recuperate a history of racial comparison that has long been submerged beneath uncomfortable silences that surround the rise of American genetics.

Agassiz's work at the Harvard Museum was inspired in part by the second site of my prehistory, a branch of experimental physiology developed at the Humboldt University of Berlin in the mid-1800s. Johannes Müller, the other grandfather of my book, taught anatomy and physiology at the university during the mid-nineteenth century. While he was in Berlin, Müller developed new techniques in experimental science that made it possible to observe the motions inside an animal's body in the moments before it died, using a practice called vivisection. His innovations in experimental physiology allowed scientists to understand the body's working, through evidence based on direct observation of animals' organs. The result was a new model of quantitative laboratory knowledge, inspired by the startling possibility of quantifying invisible vibrations of hearts, lungs, and nerves through a complex series of mechanical tools in the laboratory that transformed moving bodies into graphic images. Müller's students (who included Hermann von Helmholtz, Emil du Bois-Reymond, Carl Ludwig, and Ernst Haeckel) adapted the tools of physiology to ask new questions in psychology, investigating the workings of nerves, ears, eyes, and even of beauty itself. By the end of the nineteenth century, the tools of Berlin physiology had been appropriated into a host of new applications that included comparative studies of music.

While the genealogy of Agassiz is a lineage of people, my book's genealogy of Müller is a lineage of devices. Existing histories of audio technology describe the phonograph's emergence from studies of the human voice conducted in the late 1800s.[10] This book maps the ethics and practices of those studies backward, into the valuation of animal bodies in the 1850s. The same tools that were originally used to transform vibrations from within the bodies of dying laboratory animals into images were adapted over time to transform the vibrations of sound into graphic inscriptions in studies of music. The devices borrowed from vivisection ranged from the ear phonautograph that Edison built in 1874 using a human ear to the sound spectrograph developed in the 1950s, a tool that is still used today in scientific studies of animal vocalizations.[11] Considered together, these tools and the graphic scores they create took a set of standards that were originally developed in response to the practice of animal vivisection, and exported those standards into the study of sound as the criteria for objectivity. Today, such graphic tools continue to tie constructions of sonic knowledge to the ethics and practices of animal laboratory research, constructing "sonic" objectivity.

The importance of European laboratory science to music's culture has been well documented in studies that draw connections between scientific thought and twentieth-century music. Historians have shown the impact of acoustic physics, phenomenology, and aesthetic philosophies on the musical epistemologies of the twentieth century.[12] In my book, I turn from those histories of ideas to a history of bodies, in order to consider more closely how tools that were developed in response to the bodies of animals came to be appropriated for the objective study of sound. This tradition is an important prehistory for the material I present in my chapters, where I will argue that the valuation of animals in science served as the foundation upon which later evaluations of sonic culture were built. This approach connects music's epistemologies to an overlooked history of bodily evaluations, in which the classification of different types of music was interwoven with the exclusionary culture in which knowledge was built upon a privileged relationship between European scientists, animals, and other "others."

This is not just a story about music scholarship's relation to scientific methods, but about the centrality of animal bodies and voices to an empirical ethics that was applied to music-making bodies of every kind, and that continues to surface today in the standards that affect how music is classified and catalogued. The scientific traditions I explore in this book are often exclusionary, dependent on colonial, gendered, and racial economies that privileged white European masculinity while relying on the labor and lives of those who fell within a catchall of difference. The work of white women, nonnormative white men, nonwhite people, and nonhuman animals was often marked within these traditions as menial or illegible. In cases where exclusion was central to the process of scientific research, it also affected the way questions about music were asked and answered. The protagonists of my chapters operate within the borderlands of this tradition, and their personal lives and careers were conditioned by the structures of this unequal economy. Within those structures lie many nuances. Privileged individuals can be victims as well as perpetrators, and good science can reveal real truths that emerge as fiction within the confines of interpretation. The trajectory of my book is defined by the quandaries that these individuals discovered, described, or repressed within this exclusionary history as they confronted the conditions in which animals represented broader categorical differences that included races, nations, genders, and cultures as well as species.

The histories of musical comparison and evaluation that emerge from these two traditions raise important questions about the broader genesis of "the humanities" as we know them. My book's narrative implicates the category of "the animal" as a point of departure for turn-of-the-century musicological research that was not strictly humanistic, but relied on perceived connections between the status of the animal and race, gender, and other forms of naturalized difference. On the surface, today's construction of music as a humanistic endeavor seems unrelated to these older categories. In this book, however, I argue that contemporary humanism developed as a corrective measure amid a backlash against racial evolutionism that occurred after World War II. While comparisons between animals and human races were rejected in the aftermath of the war, the category of "the animal" itself remained unchanged, and the capacity of the animal to facilitate universalizing categories of difference remained unexamined. The postwar construction of the humanities as divided from science and species has only intensified the singularity of "the animal" as a foundational category of difference. In the pages of my book, I focus on stories in which animals have constructed modern sonic culture. But I also seek to make present, through these stories, the missing histories of race and difference that underpin the modern construction of the humanities as we know them.

## A TASTE FOR RUPTURE

By placing the divisions between humans and other animals at the center of my narrative, I want this book to be an intervention in the kinds of questions we ask of humanism and the humanities. What is an anti-speciesist history of music? What is the singular of the plural entity called the humanities, and what is its purpose? What happens if you believe, as I do, that the question "What is a humanity?" is inseparable from histories of difference? How does one use the disciplinary engine of knowledge we call the humanities when that engine is predicated on deeply unequal relationships between categories of difference such as race, gender, culture, or species? This intervention is not a posthumanist one, for posthumanism is tied to the history I explore. Instead, two questions guide my study of music's taxonomies:

What is difference?

What is a humanity?

What is difference? To paraphrase animal studies scholar Una Chaudhuri, thinking about animals is reshaping the way we approach cultural categories in every corner of academic discourse: gender, class, race, nation, age, profession, sexual orientation, marital status, and, of course, species.[13] Today a growing literature addresses the intersection of animality with historical categories of gender, race, sexuality, class, and nation.[14] That intersection of human and animal difference is one I find particularly salient in the modern history of Western music, where divisions between genres and styles have become intertwined with narratives of cultivated identity.

The measure of difference that compares a sparrow's song to a Beethoven symphony is also, I argue, a practice that preconditions our present experience of culture, race, and biology. Within the context of my book, I treat difference as a particular heritage of the late nineteenth and early twentieth centuries that has affected the politics and structure of music's orderings in the early twenty-first century. In my book's stories, difference was often treated by those who studied songs as if it were a unifying, natural principle that connected existing categories such as race, nation, or gender to a broader state. Within the lives of the actors that occupy my narrative, being a woman, or an immigrant, or an Italian, or a dog were all unique, particular experiences. But while such experiences have unique histories and contexts, the practice of comparing songs across racial, national, gendered, and species boundaries demanded a notion of difference as a foundational category premised on the belief that forms of difference exist in a shared state of diversification waiting to be compared, analyzed, and organized. In the cases that I explore, animals were the exemplar of that state. For those who have lived with and through this epistemology in which a sparrow and Beethoven sing on a shared musical taxonomic plane, to talk about birds is to talk about race, gender, sexuality, or class; and to talk about gender, race, or class is to talk about species.

This brings me to my second question, "What is a humanity?" For those who believe that modern epistemology has been shaped by the notion that all forms of difference implicate one another, to ask "What is a humanity?" is to ask about more than the divide between human and nonhuman actors. Scholars have had a taste for the rupture between the humanities and the sciences since the late twentieth century, when the divisions between humans and animals, laboratories and libraries, and genetic inheritance and cultural heritage captured their interest.[15]

In the twenty-first century, posthumanism, new materialism, and object-oriented ontology have built on this foundation to bring forth rich theories about nonhuman agency and the interpretation of objects. My book's exploration of musical taxonomy is a kind of counter-history of this movement toward the nonhuman. In that counter-history, I figure the question "What is a humanity?"—the act of recognizing the problematic division between the human and nonhuman—as an act that is impossible to perform without partial or full participation in the past century of taxonomic interventions that are always already raced, gendered, sexed, animaled, and otherwise ordered through difference.

What is this book *not*? It is not a comprehensive, consecutive history, but a speculative one. Its chapters are arranged chronologically and have substantial interconnections, but each chapter can be read independently. I have tried to fill these chapters with events and experiences that are inherently interesting and meaningful: a young woman's discovery of a new species in Angola; a deaf naturalist obsessed with birdsong; a Berlin psychologist trying to perform vivisections on songs; a British biologist tracing pictures of animal music. The protagonists of these chapters are loosely related by their shared teachers, institutions, and disciplinary norms. The history that results is not a critique or rebuttal of the posthuman, but an alternative. That alternative is limited to about a century and a half, and it is within that period that I locate words such as "modern," "humanity," or "animal." This book is also not in favor of dispensing with notions of difference or similarity, but seeks to provoke questions about how such ideas can be used responsibly. Songs have a unique place in this history, and I use them to suggest a much broader narrative than one book can truly contain.

My questions "What is difference?" and "What is a humanity?" are premised on the expectation that the unsustainable divide interrogated by posthumanism cannot be resolved without simultaneously addressing the question of how difference has been historically constructed in the twentieth century. I seek to make a beginning toward that effort in the pages that follow. In the conclusion of this book, I call those aspirations the "animanities," an outgrowth of my experience with posthumanism and animal studies in their aesthetic and academic forms, situated in the study of culture as a more-than-human reality.

What is difference, and what is a humanity? The stakes of these questions range from the issue of who does interdisciplinary work and how, to the financial pressure to leverage cultural inquiry into scientific funding sources in the twenty-first century. But knowledge is also at stake, and I will argue in the chapters to come that how we make knowledge about culture and music is intricately tied up with the question of how we evaluate difference, and how that evaluation is also a *valuation* of different lives.

## AMNESIA

If humanism does not answer the needs of this book, what does? George Santayana once wrote, "Those who cannot remember the past are condemned to repeat it."[16] To remember the past is very much the project of this book. I begin that process of remembrance by remembering the birds' voices that gave the original context to Santayana's words: "Progress, far from consisting in change, depends on retentiveness . . . when experience is not retained, as among savages, infancy is perpetual. Those who cannot remember the past are condemned to repeat it . . . Thus old age is as forgetful as youth, and more incorrigible; its memory becomes self-repeating and degenerates into an instinctive reaction, like a bird's chirp."[17] Santayana's *The Life of Reason, or the Phases of Human Progress*, published in 1905, shared his generation's conviction that animal nature had much to reveal about the capacity to speak and, therefore, to be heard. The chirping bird that occupies the lowest place in Santayana's taxonomy of memory was put into print only a year after Mathews compared the song sparrow to Verdi's "La donna è mobile," and was followed by many more such comparisons.

In the intervening century, the history that connects music's taxonomies to the status of animals has largely faded and been forgotten. Yet interconnections between the lives of animals and today's constructions of race, gender, sexuality, and other forms of difference have been thoroughly demonstrated by recent scholars within animal studies.[18] By recuperating the history that connects animal lives to measures of musical difference, my book is meant as an intervention within the postmodern, postwar, posthuman construction of humanism and the humanities as spaces in which animality and other constructions of difference can be treated separately and ahistorically. By framing my history of music's taxonomies as a counter-history of the present, I hope that my readers adopt an active role in reimagining music's orders outside of established human/nonhuman, culture/nature, humanities/science dyads.

The first half of this book engages the period between 1871 and 1950 to show how metaphors of animal life and death framed a process of collecting, classifying, and analyzing songs as a measure of difference. I begin chapter 1 at the end of the nineteenth century, with a print war between Charles Darwin and Herbert Spencer about the evolutionary origins of musical aesthetics. Darwin's

*The Descent of Man, and Selection in Relation to Sex*, published in 1871, proffered a controversial theory that aesthetics and music originated in artistic, skillful, and self-aware animal minds. The book sparked heated responses as thinkers in the late nineteenth and early twentieth centuries asked whether other animals, particularly birds, possessed the capacity for culture. In the ensuing comparisons between animal and human song, musical evolution became a means to debate the right to personhood.

The following three chapters follow the course of music's place in hierarchies of difference through traditions of collecting, classifying, and analyzing songs. In chapter 2, I explore the introduction of collected songs into institutional collections as the musical equivalent of natural history specimens. These institutional song collections offer a revealing glimpse of the way biological models of identity and identification shaped beliefs about songs and those who produced them. In chapter 3, I turn to the experiences of the professional song collectors who made institutional collections possible. In this chapter I shift from institutional history to biographical narrative, in order to show how notions of identity and difference were intertwined with deeply personal experiences of inequality in the profession of song collecting. I focus in this chapter on the experiences of two collectors from the 1920s and '30s, a psychologist and zoologist named Wallace Craig, and a music expert named Laura Craytor Boulton. Through their stories, I examine the way essentialized beliefs about bodily inequality became the conditions under which song collectors crafted both their collections and their careers.

Turning from the collection and classification of songs to a music laboratory in Berlin in the first half of the twentieth century, in chapter 4 I explore the ways that notions of musical data in the 1920s and '30s were premised on a nineteenth-century ethics of vivisection, in which animal lives were traded for empirical knowledge, taking the form of graphic inscriptions made by the tools of the physiological laboratory such as the kymograph. Musical data, I therefore suggest, was the written result of this exchange of difference. Instead of answering the question posed by the tradition of animal vivisection—"Is knowledge worth killing for?"—I ask my readers to resist the premise of the original question, which posits that our job is to weigh the value of life against the value of knowledge.

Chapter 5 marks a turning point in my narrative, in which human identity was redefined after World War II through a separation of the spheres of biology and culture. The second half of my book explores this post-1950 period, starting with the crisis in scientific knowledge that laid the groundwork for this changing world. I show here how the dramatic rejection of racial comparison in evolutionary biology and the social sciences made it desirable for studies of music to become a separate topic from studies of evolutionary development. In chapter 5, I explore the reinvention of evolutionary theory through a distinction between genetic and cultural inheritance. Paying special attention to the rise of genetics at the laboratories of Cold Spring Harbor, I suggest in this chapter that the special status of the human after World War II—the division of culture from biology that defined postwar notions of the humanities—is the postmodern condition of race.

In chapter 6, I show how new audio technology offered a welcome pretext in this divided environment to replace the musicianship of the prewar "sonic specimen" with machines that would dramatically change perceptions of animal "music," making it the mark of a neurological, rather than a cultural, capacity. As a result of this changing practice, being musical in the post–World War II world meant, by definition, being human, while studies of birdsong and other animal vocalizations fell under the rubric of a laboratory science divorced from conventional ideas of "music." In this chapter, I trace that rupture of knowledge into the disparate worlds of human song studies and studies of birdsong that postdate 1950. The result is a dynamic difference in disciplinary standards for evidence, exemplified by the role of the objective graphic image, first explored in chapter 4 as the picture of vivisection.

The closing chapter of my book revisits traditions of the sonic specimen and aural identity in the context of audio field guides to birdsong. These are objects that both replicate and challenge the postmodern divorce between birdsong and human music. In this chapter, I focus on three exceptional cases, in which listening to other species is a guide toward alternate notions of the self located in exotic paradises. Olivier Messiaen's *Catalogue d'oiseaux*, Steven Feld's *Rainforest Soundwalks*, and Miyoko Chu's *Birdscapes* offer a shared yearning for something best described as the Garden of Eden. These cases conclude my book with three imagined paradises that respond to a world that has been partially determined by the restriction of things of the spirit to the human experience. I end the chapter with questions about how we might imagine paradise beyond the restrictions of the postwar, postmodern, posthuman division between human and nonhuman life.

In the course of these seven chapters, seven themes emerge that offer a foothold for rethinking the more-than-human conditions of culture that emerge from this history. In the conclusion of this book, I imagine these themes as the foundation of the practice I call the animanities, the more-than-human study of culture. I chart a path into the animanities through the stakes raised by each chapter—personhood, identity, difference, knowledge, postmodern humanity, subjectivity, and paradise—as they lay claim to this imagined field of inquiry. Neither quite posthuman nor merely animal, my mapping of the animanities closes with an assertion of the urgency of considering these stakes in a world where human identity itself is being remade at all levels, from our dependence on a rapidly changing environment to the medically altered molecules of our DNA.

The ties between culture, race, and species are of more interest today than they have been for many years. Since 2000, the emergence of brain imaging technology, the completion of the Human Genome Project in 2003, the discovery of Paleolithic bone flutes in 1995 and 2008, and the use of CRISPR technologies to edit human genes in 2013 have returned cultural evolutionism to the sphere of public inquiry. In the past fifteen years, science writers and music scholars including Steven Mithen, Aniruddh Patel, Steven Pinker, Edward O. Wilson, Ian Cross, Daniel Levitin, and Gary Tomlinson have published a wave of new texts on music's evolution. These contributions come at a moment when the relatively strong funding available in the sciences incentivizes science-oriented readings of music. Yet the turn toward musical evolutionism reinscribes the notion that humanistic labor is inherent to a "normal" human body, and therefore needs neither resources nor support to persist, so long as the normal human body survives.

This return to evolutionary thought calls to mind Santayana's admonition that we are condemned to repeat the past that we forget. For many, the events of the early twenty-first century echo the crises of nationalism, immigration, global violence, and technological change that shaped the first half of the twentieth century. Amid those echoes, the lives of animals point toward what we have forgotten: something about how constructions of difference are the mechanism of posthumanism as well as its material; something about how music came to be removed from animal voices; something that we cannot quite remember about the nature and politics of hearing identity and difference.

To borrow from Santayana, I use the following pages to seek the empowerment offered by history's memory. Though my book's narrative begins at the end of the nineteenth century, its place is in the present, a moment at which musical classification, evolutionary science, and commercial metadata have converged. This project does not seek to erase, eradicate, or fully explain the differences between humanism and scientific inquiry. But it does seek to bridge those differences by articulating one of the historical causes that has restricted music to the human sphere: the persistent power of culture to categorize difference, and to determine who it is that counts as a person.

# PERSONHOOD

———

The first chapter of this book surveys a series of debates
about animal musicality that unfolded in print during the late
nineteenth and early twentieth centuries in Europe, Great Britain, and
the United States. Competing theories of evolution structured these
debates around assessments of sentience. By examining the process
by which evolutionary theories of music were tied to assessments of
interiority, this chapter outlines a crucial context in which later studies
of animal musicality occurred. At stake in these assessments was the
right to personhood, for the debate about animal musicality was a
debate about who was, and was not, a person.

# ONE

# Why Do Birds Sing?
# And Other Tales

In 1913, Henry Oldys, a biologist working for the US Department of Agriculture, wrote enthusiastically to readers of the nation's premier journal of bird science, *The Auk*, "Astonishing and revolutionary as it may seem, there is no escape from the conclusion that the evolution of bird music independently parallels the evolution of human music."[1] Born in 1859, the year Charles Darwin published *On the Origin of Species*, Oldys was part of a generation exposed to controversial new ideas about the role animals played in human social and cultural development. In the foreground of this controversy were the names of Darwin and Herbert Spencer, whose ideas about the animal origins of human song put birdsong on the map of the new science of cultural evolution. Over the course of the next century, thinkers like Oldys increasingly took up the question of how animals, especially birds, were tied to the evolutionary origins of song. As music historian Robert Lach explained, the question "Why do birds sing?" was seen by scholars of culture and science as "the key to the problem of the origin of language and music."[2] At its heart, this was a question that meant rethinking the capacity for music, raising surprisingly complex questions not only about animals, but about human beings and just how special their musical abilities really were.

Why *do* birds sing? By asking this question at all, intellectuals such as Lach and Oldys imagined a capacity for music that held forth the tantalizing promise of connecting song, still imagined today as deeply human, to a totally alien world of nonhuman experience. The evolutionary discourse that developed in the late nineteenth and early twentieth centuries around Darwin's and Spencer's texts debated the relation of these two worlds, struggling to fill the gap between the music of modern civilized humans and the primal sounds of their animal ancestors. The unknown space of this gap contained the key to a biological science of culture, in an era deeply invested in justifying racial hierarchies through science. As evolutionary historian Peter Bowler has explained, "virtually all evolutionists [at this time] accepted the linear image of human origins and used it to justify prevailing racial prejudice."[3] It is a period in which Darwinian evolution coexisted with diverse theories of saltation, orthogenesis, neo-Lamarckism, eugenics, social Darwinism, and other approaches that invited comparison between biology and culture.[4] Against this backdrop, music emerged as one of the "missing links" that promised to fill the gap between biology and culture in evolution's story with the sounds of animals' cries and "primitive" human songs.

At the center of these debates about music were the works of Charles Darwin and Herbert Spencer. The two men were well known, Darwin for his history of biological evolution in the *Origin of Species* and Spencer for his studies of human development in works such as *The Principles of Sociology*. Their theories of music reflected their broader interests: Darwin, the biological historian, argued that music like birdsong affected mate selection and offspring, while Spencer, the social historian, argued that song had to do with the boundary between human reason and emotion.[5] For Darwin, birdsong and human song were both legitimate forms of music; for Spencer, only humans made true music. As later authors debated music's place in evolution, the subtleties of Darwin's and Spencer's arguments were sometimes lost in accounts of Darwin as the defender of animal musicality, and Spencer as his opponent.

This discourse took shape across numerous disciplines, sometimes in indirect ways. In this chapter, I reconstruct its main arguments by drawing on voices from a wide range of disciplinary identities and a period spanning several decades, connecting the threads that crossed these disciplinary and temporal divides. The Darwin-Spencer debates about music's origins mark an initial stage in this discourse, which took place in the late nineteenth century. Its primary contributors were European evolutionists, including the biologists August Weismann and George Romanes, and the British psychologists James Sully and Edmund Gurney. Other voices in these debates, like Lach, were formally trained music scholars from Europe or the United States. They were the first generation of "musicologists," historians of music who turned away from biographies of great masters like Beethoven and Bach toward a broader-reaching social science inspired by figures such as Spencer.[6] A third set of voices in musical evolutionism came from the natural and social sciences, where biologists, psychologists, and anthropologists had an interest in connecting animal aesthetics to human development, particularly in Britain and the United States, where Darwinism was on the rise. Finally, the voice of the naturalist had an important role in these debates as well, contributing firsthand experience shaped by hunting, hiking, and observation. This was particularly true for experts in birdsong, a field so new that there was no formal schooling in it—Oldys was a case in point, working as a lawyer and auditor before building his reputation as a biologist with the Department of Agriculture.

Although many of these men and women operated in separate professional spheres, they were connected by books, pamphlets, and journals. Print was the medium of their discourse, allowing widely flung experts and amateurs to trade ideas. In the pages that follow, I trace the war in print over animal musicality from its initial phase in the Darwin-Spencer debates of the 1870s to later appropriations of their ideas in the early twentieth century. Although Darwin's and Spencer's opposed theories of music were not solely about music's role in determining human uniqueness, the texts inspired by them often returned to this refrain. In tracing this burgeoning science of music to the texts that inspired it, I hope to show how listening to culture and listening to nature merged over a period of several decades to produce a practice of hearing biocultural difference, where song became a measure of other species' worth. The stakes of this debate were not merely an argument about evolutionary origins. They were about personhood, for to be a musician—human or animal—was to be a person.

## BIRDS IN PRINT

Though Darwin and Spencer were a touchstone for debates about animal musicality, interest in the topic of animal musicality was already in the air. Scholars of the nineteenth century published anecdotal accounts of dogs, cats, and even horses barking, yowling, and marching in time to human music, hoping to understand where to draw the line between human and nonhuman ability.[7] In the early twentieth century, psychologists and physiologists published measurements of animals' pulse rates in response to music, and the salivation of dogs as they recognized melody, harmony, and tempo.[8] Decades before the arrival of "ecomusicology," George Herzog was able to ask members of the American Musicological Society, "Do Animals Have Music?" while The Musical Quarterly surveyed the musical taste of a veritable zoo of animals including dogs, cats, birds, snakes, monkeys, mice, cows, horses, chickens, and a flying squirrel.[9] The possibility that music was a universal capacity remained open well into the twentieth century:

as Herzog put it, "there seems to be no criterion for any theoretical separation of the vocal expression of animals from human music."[10]

Advocates of animal musicality could be quite persuasive.[11] In 1871, the year Darwin's *Descent* was published, an article by American minister Samuel Lockwood in *The American Naturalist* sparked a decade-long vogue in mouse music. Lockwood's essay documented his singing pet mouse Hespie in astounding detail. Describing Hespie's daily life and singing habits, Lockwood recorded her high-pitched songs using prose descriptions and transcriptions of Western musical notation. Arias like her "Wheel Song," he argued, were testaments to her musical taste, precision, and baroque sensibility for ornamentation.[12] Interest in Lockwood's singing mouse spread from Darwin's *Descent* to the British journal *Nature*.[13] Another symptom of animal music's popularity was British nature writer Charles Cornish's *Life at the Zoo: Notes and Traditions of the Regent's Park Gardens*, published in 1894. The book documented a series of informal experiments in which Cornish engaged a violinist to play for animals at the Regent's Park Zoological Gardens in London.[14] Cornish, an author of British nature guides, walked around the zoo with his assistant playing popular tunes for selected animals, usually the Scottish reel "The Keel Row." (Scottish tunes were still considered exotic "natural melodies" free of any preconceived system, and thus might have seemed more appropriate for animal ears.)[15] The results of Cornish's "experiments" were occasionally absurd—he claimed, for example, that wolves and sheep responded differently to "The Keel Row" because they were natural enemies, and that sheep preferred the "Shepherd's Call" sequence from *William Tell* because they were pastoral animals. But many of Cornish's examples were compelling, such as the photograph that showed axis deer turning their large ears to listen to the violinist.[16] Like more and more descriptions of musical animals, Cornish's account documented animals who were sensitive and aware musical listeners.

Of all these anecdotes of musical creatures, those describing birds were the most compelling. "From the wearisome sameness of a sparrow's chirp," wrote the British psychologist James Sully, "up to the elaborate song of the skylark or nightingale, there presents itself something like a complete evolution of vocal melody."[17] The variety that separated simple one-note calls from the long, complex songs of birds like the nightingale seemed to contain the story of music's development. Music scholars, ornithologists, naturalists, and science writers looked at topics ranging from the evolution of musical taste in birds to comparisons between American birds and American composer Stephen Foster's songwriting.[18]

## NO. I. THE WHEEL SONG.

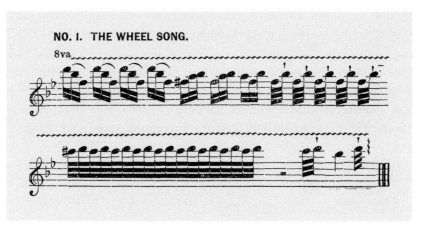

FIGURE 1.1 Hespie the mouse's favorite song in the exercise wheel.
Lockwood, "A Singing Hesperomys."

FIGURE 1.2 Axis deer listening to "The Keel Row."
Cornish, *Life at the Zoo*, 118.

Musical notation in bird guides reinforced the impression that birdsong was a musical artifact, subject to the laws of rhythm, melody, and harmony. Asserting the fundamental musicality of birdsong, ornithologist Francis Allen asked, "how can we escape imputing the origin and development of this beauty in bird-song to an aesthetic sense in the birds themselves?"[19] Perhaps most important, birds held a place of honor in Darwin's claims about music, for he argued that one could "hear daily in the singing of birds" evidence that animals uttered musical notes.[20]

## DARWIN AND SPENCER

> There are, as is well known, two leading theories with regard to the origin
> of music—Mr. Darwin's and Mr. Herbert Spencer's.[21]

Until the end of the nineteenth century, discussions of music's natural origins ranged from the mind-bending powers of musical ratios in Plato's *Republic* to Rousseau's *Dictionnaire de Musique*, completed in 1767. As early as 1650, Jesuit scholar Athanasius Kircher described a remarkable sloth singing a diatonic hexachord, as well as a bevy of singing birds that made later appearances in many a discussion of nonhuman musicality.[22] Thinkers and philosophers such as Rousseau and Kircher, and the more recently minted names of Schiller, Herder, Wagner, and Helmholtz, added weight and expertise to questions about the origins of music throughout the nineteenth century and into the twentieth. By the end of the nineteenth century, music had become the most prestigious of the fine arts, its unique powers of emotional expression making it, in the words of British social philosopher Herbert Spencer, "one of the characteristics of our age."[23] Music was repeatedly described as the "language of the soul" by intellectuals ranging from Johann Herder to Russian author Andrei Bely.[24] But during the late nineteenth and early twentieth centuries, music's spiritual language increasingly took on an evolutionary tone, becoming integral to a growing discourse about biology's relationship to human culture and identity. Even the fictional detective-violinist Sherlock Holmes became a musical evolutionist, reminding his companion Dr. Watson that daring ideas about music and language were necessary to grasp the breadth of nature's ways:

> "Do you remember what Darwin says about music? He claims that the power
> of producing and appreciating it existed among the human race long before
> the power of speech was arrived at."

"That's a rather broad idea," I remarked.

"One's ideas must be as broad as Nature if they are to interpret Nature," he answered.[25]

By the time Sherlock Holmes was hunting down criminals with his cold British logic, Charles Darwin and Herbert Spencer had become public voices of modern science, and the two leading figures in musical evolutionism: "the two theories of the origin of music which have claimed the attention of critics are Darwin's and Spencer's."[26]

Framing these ideas "as broad as Nature" was a much older set of beliefs about the place of language in human identity. In Darwin's day, it was widely assumed that language demarcated a definitive leap between human and animal psyches, a rupture whose relation to music was uncertain. Eighteenth-century figures such as Rousseau and Herder had popularized theories that language was the seat of human intelligence, emphasizing the rational ties that bound *Homo sapiens*, the man of reason, into a universal talking brotherhood. Enlightenment theories of language prompted so much speculation over the following century that the Parisian Société de Linguistique issued an ineffectual ban on the topic of language origins in 1866, hoping to control further debate.[27] Instead, the topic continued to thrive, fostering guesses about syntax, grammar, and translatability as possible trademarks of the intelligent mind separating human language from animal vocalizations. As Henri Bergson explained in *L'Évolution créatrice* of 1907, human language hinged on the distinction between instinct and invention: "*the instinctive sign* [of the animal] *is* adherent, *the intelligent sign* [of the human] *is* mobile"—that is, animal cries had a fixed, instinctive meaning dictated by nature, while human words were a matter of invented convention and, therefore, a mark of superior intelligence.[28] Even with the rise of colonialism and a corresponding emphasis on racial distinctions at the turn of the century, Bergson's contemporaries still believed, as Herder had, that humanity could be universally defined as "the talking animal."[29] Attempts like Richard Garner's to decode primate language, described by historian of science Gregory Radick, extended this intelligent sign no further than humanity's closest relatives.[30]

Music had a very ambiguous relationship to this view of language. The sounds of birdsong and other musical animal vocalizations were not the intelligent signs of language, nor were they the proto-linguistic responses biologists dubbed "calls." But neither were they innate—some animals had to learn their songs through such arduous study that the idea of animal conservatories with "properly qualified professors" to refine animals' musical abilities was floated at the turn of the century.[31] Though not a mark of reason, song in animals seemed very similar to song in humans. As a result, music occupied an undefined—and potentially contested—territory between animal expression and human linguistic rationality. For evolutionists like Darwin and Spencer, defining this ambiguous territory would mean defining what it meant to be human.

Darwin's *The Descent of Man, and Selection in Relation to Sex*, published in 1871, used music and aesthetics to challenge prevailing beliefs about this boundary line between humans and other animals. "The place of song in the life of the bird has since the days of Darwin been a question of dispute between the scientists," wrote ornithologist Chauncey Hawkins in 1918.[32] This "dispute" echoed the broader anxieties about human uniqueness that had surrounded the *Origin of Species*, but focused on a new question: whether culture, instead of biology, was what set humans apart from the animals. One of the most controversial implications of the *Descent* was that it did not.

Darwin's *Origin of Species* from 1859 had located biological history in a process of natural selection that depended on a species' successful reproduction. In the *Descent*, Darwin argued that aesthetics had their origins not in divine providence but in a special kind of selection called "sexual selection," in which animals chose their mates. Darwin spent nearly a quarter of the *Descent* on birds and sexual selection, displacing primates in a surprise move that voted with ink in favor of music's role in human development. He suggested that human aesthetics descended directly from the selective processes that occurred when animals had the power to choose their own mates, arguing that generations of attracting, being chosen, and having offspring had bred a sense and appreciation of beauty in certain animals. For Darwin, these animals had the same capacity for musicality as humans. The British naturalist lauded the musical abilities of spiders, frogs, tortoises, alligators, gibbons, dogs, seals, and even Lockwood's mouse Hespie, returning throughout his text to the question of music's origins.[33] Passion, love, breeding, persuasion, and social power all had a place in this unusual narrative of humanity's link to animals through a process that looked very much like free will. "The impassioned orator, bard, or musician, when with his varied tones and cadences he excites the strongest emotions in his hearers, little suspects that he uses the same means by which his half-human ancestors long ago aroused each others' ardent passions during their courtship and rivalry."[34] Darwin's theory was popularized by figures ranging from the fictional Holmes to psychologists such as James Sully and Edmund Gurney, and musicologists such as Jules Combarieau and Robert Lach, gaining a considerable number of adherents by the early twentieth century.[35]

But sexual selection was a very different story from the cold and random chance of natural selection. Many intellectuals, however respectful of Darwinian evolution, were troubled by the theory that aesthetics' origins lay in the loves, choices, and passions of animals. For many, it was simply too difficult to reconcile the impersonal forces of nature that dominated evolutionary theory with the personal, fickle individuals *creating* emotions that led to *choosing* a mate whose offspring would then *choose* future generations. The first-ever book on birdsong's evolution, Charles Witchell's *The Evolution of Bird-Song, with Observances on the Influence of Heredity and Imitation* of 1896, devoted almost all of its opening chapter to a critique of Darwin's theory of musical evolution, concluding that despite certain compelling features of sexual selection, the songs of birds could not "reasonably be considered to be directly occasioned by the emotion of love."[36] The biologist Alfred Russel Wallace, who had independently engineered his own natural selection theory, argued that Darwinian sexual selection in which animals seemed to "consciously" choose their future was wrong, and should be replaced with natural selection.[37] Wallace's colleague August Weismann agreed, writing that "the musical sense is not a result of sexual selection," and arguing instead for a nineteenth-century version of Steven Pinker's description of music as deliciously unnecessary "auditory cheesecake"—a delightful coincidence of acoustic development useless in survival terms.[38]

Darwin's most visible opponent on the matter of musical evolution was his contemporary Herbert Spencer. Spencer, a British social and political philosopher, had been a public advocate for Darwinian natural selection. Both Darwin and Spencer thought music might offer new ways to understand human differences. But while Darwin's work was rooted in theories of biological reproduction, Spencer's approach was framed by sociology. His writings about music, which located aesthetics in human social life rather than biological history, became the public foil for Darwin's thinking on aesthetics, with "Mr. Darwin's and Mr. Herbert Spencer's" theories becoming polar twins in the discourse of music's origins.[39]

The first of Spencer's texts on music actually predated the *Descent*, becoming newly relevant after Darwin's controversial book was published. Written in 1857, two years before the publication of *On the Origin of Species*, Spencer's "The Origin and Function of Music" put forward an argument very similar to Herder's from the previous century, treating music as the emotional component of early language. Spencer approached all animal sounds as exclamations or "impassioned utterances" caused by restless emotional energy. Birdsong was, in his view, functionally equivalent to barking dogs, purring cats, roaring lions, and any other shrieks and moans of pain, joy, or suffering.[40] Human vocalizations, on the other hand, were uniquely rational, developing into speech, chant, and song. The evolution of these utterances could be explained as a side effect of environmental pressures and local competition, but had nothing to do with choice, aesthetics, or, especially, sexual reproduction. Spencer's take on music was fundamentally a sociological one, connecting aesthetics to prevailing beliefs about reason, language, and the emotions. It was about human experience, not biological history; Spencer's theory was entirely disinterested in the development of the syrinxes, tongues, mouths, and ears that accompanied song.

Darwin had briefly referenced Spencer's essay in the *Descent*, but he provided a more vigorous response in *The Expression of the Emotions in Man and Animals* (1872). Though respectful of Spencer's general ideas, Darwin argued that in treating all animal sounds as the same type of sound, Spencer and those like him had essentially missed the point. Spencer's pig uttered generic exclamatory grunts; Darwin's pig uttered a special "deep grunt of satisfaction . . . when pleased with its food, [that] is widely different from its harsh scream of pain or terror."[41] The barking dog that opened Spencer's essay had yet another way of making sounds: "But with the dog . . . the bark of anger and that of joy are sounds which by no means stand in opposition to each other; and so it is in some cases."[42] Spencer and other advocates of the natural-selection approach to music, Darwin implied, simply failed to recognize musicality as a distinct mode of sound production, thereby failing to recognize what everybody else knew: "that animals utter musical notes is familiar to every one, as we may hear daily in the singing of birds."[43] Psychologist Edmund Gurney's *The Power of Sound* (1880) added further rankle to Darwin's criticisms, claiming a Darwinian approach was "directly opposed" to Spencer's theories.[44] Gurney even parodied Spencer's idea that melody originated in speech cadences, with a mock example tracing J. S. Bach's cantata *Ein' feste Burg ist unser Gott* to pseudo-evolutionary origins in the English exclamation "Heigh-ho!"[45]

In response to his critics, Spencer had his essay reissued in 1890 with a special postscript responding to Darwin and Gurney. "Mr. Darwin's observations are inadequate, and his reasonings upon them inconclusive. Swayed by his doctrine of sexual selection, he has leaned towards the view that music had its origins in the expression of amatory feeling."[46] "Mr. Gurney," in his turn, suffered from "deficient knowledge of the laws of evolution" in his support for Darwin.[47] Objecting to the suspicious developmental gap between birds and primates in Darwin's explanation, Spencer attacked the idea of birdsong as a courtship ritual, calling it "untenable" and adding, "What then is the true interpretation? Simply that like the whistling and humming of tunes by boys and men, the singing of birds results from overflow of energy."[48] For Spencer, such sounds were merely precursors to expressions of emotion that would eventually become the cadences of speech and song. In the ensuing decade, Spencer reprinted his essay with further additions at least six more times before his death, cementing the perceived polarity between his work and Darwin's. It was an opposition that generated a broad audience for musical evolutionism, building a tenuous musical bridge across the gap between biological evolution and human history.

## SOCIAL EVOLUTION

Although the Darwin-Spencer debates appeared to be about sexual reproduction, in practice they were often about social evolution.[49] For intellectuals like Sully, "the genesis of animal music is at the same time the explanation of the early developments of song in the human race."[50] In this approach, the songs of primitive human beings could be compared to the songs of advanced animal species. The result was, as science writer Grant Allen put it, a map of aesthetic evolution "from the simple and narrow feelings of the savage or the child to the full and expansive aesthetic catholicity of the cultivated adult."[51] Bird fanciers had long made comparisons between chicks raised by foster parents learning the foster species' song, and human children learning the tongue of adoptive parents.[52] References to Darwin's and Spencer's ideas became a way to channel such comparisons into an explicitly evolutionary, linear, hierarchical discourse about the relationship between birds, humans, races, and the forms of difference that lay between.

Evolutionary historian Peter Bowler has called social evolutionism "a system of cosmic progress," while Timothy Ingold names it "the *telos* of an embodied purpose," an inevitable forward-moving arrow of change that ranked race and species in order of development.[53] The hierarchy of this arrow was permeated by the nationalist politics and colonial economies of nineteenth-century Europe, Britain, and America, becoming the dominant vehicle of approaches to biological evolutionism.[54] Assumptions about race, class, and gender permeated musical explanations of human and animal difference. German evolutionist August Weismann compared the gap separating Beethoven from a "primitive" human musician to the distance between a parrot and a human, arguing that "we cannot suppose that any Beethovens were concealed among primitive men, or are running around among contemporary Australians or negroes," because "savages are lower in mental development than civilized man."[55] American naturalist Henry Oldys praised birds by connecting them to the very same savages, arguing that "the songs of some birds must be ranked above the best music of many primitive races of today."[56] Even Cornish, in his lighthearted account of zoo animals, implied familiar racial stereotypes when he contrasted the European wolf's "hideous sneer" on hearing music with the more "extreme and abject fear" of its Indian cousin.[57]

Social evolutionism's racial stereotyping was compelling in part because it addressed a real need for historians and social scientists. Older historians such as Thomas Carlyle had argued that social change was the work of influential kings and geniuses, the "great man" theory of history.[58] Biography, the main method of this approach, did little to explain changing communities or cultures; and by the end of the nineteenth century intellectuals ranging from Marx to Darwin sought alternative ways of understanding social change. While Darwin and many of his contemporaries espoused a social theory of groups rather than individuals, one of the most recognized theorists of this approach was Herbert Spencer. Spencer's views on music were one facet of this larger sociological picture. Spencerian social evolution substituted an "enormous aggregate of forces that have been co-operating for ages" for older biographical historiography, replacing the "great men" of historians like Thomas Carlyle with social science.[59] When Spencer asked his readers to understand music by comparing the sounds of a gentleman to those of a clown, or a refined lady to her servant, he argued that music was a mechanism of social change, defining—and potentially changing—human social classification.[60]

These themes of social aggregates, progressive change, and classification might have seemed like established facts to many music lovers. Nineteenth-century composers such as Wagner and Busoni had viewed music as both highest in rank among the arts, and forward moving in time. Spencer too thought music was "the highest of the fine arts," an art form he believed would open the gates to universal human sympathy in due time, fulfilling the ultimate goal of social evolution's direction.[61] By the early twentieth century, modernist composers fused primitive and futurist idioms in works like Stravinsky's *The Rite of Spring* to create a musical language out of this evolutionary outlook.

Specialists in music's history were particularly interested in social evolutionism's emphasis on human development. They openly debated the relative merits of Darwin's and Spencer's theories, arguing whether speech or song came first and producing a wave of successful textbooks with names like *The Evolution of the Art of Music* and *Music, Its Laws and Evolution*.[62] Most sided with Spencer's belief in human uniqueness. Guido Adler, one of the founders of the field, wrote that music's history was "a matter of natural selection," suggesting that musical styles were subject to evolutionary forces that made the strong flourish and diminished the weak.[63] Like Spencer, Adler advocated for a shift away from "great man" biographies to a history shaped by "epochs—large and small—or according to peoples, territories, regions, cities, and schools of art . . . without special consideration given to the life and effect of individual artists who have participated in this steady development."[64] But Adler was opposed to Darwin's sexual selection theory, writing in 1911 that, "no musical artistry can be explained by monkey instincts."[65] Charles Hubert Parry, director of the Royal College of Music in London, explained the concept of music's evolution at the end of the nineteenth century in a similar language of anthropological comparison:

> The basis of all music and the very first steps in the long story of musical development are to be found in the musical utterances of the most undeveloped and unconscious types of humanity; . . . Such savages are in the same position in relation to music as the remote ancestors of the race before the story of the artistic development of music began; and through the study of the ways in which they contrive their primitive fragments of tune and rhythm, and of the principles upon which they string these together, the first steps of musical development may be traced.[66]

Parry's approach favored racial rather than species development, deriding Darwin's "childish theory" that music originated in birdsong.[67]

Yet birdsong remained a convenient point of reference for other music scholars. Adler's student Robert Lach, who supported Darwin's theory of sexual selection, developed a taxonomy of musical ornaments based on the calls and songs of birds.[68] Lach even suggested that some of the great composers of the Italian baroque had birdlike taste, turning to the strophes of Caccini, Caldera, and Castello.[69] Shortly after Lach, American composer and critic Daniel Gregory Mason opened one of the first general histories of music published in the United States with a transcribed version of a bird near his home in Massachusetts, whose adherence to a D major triad impressed the musician with the likelihood that primitive scales derived from a natural order perceptible to the other animal kingdoms.[70]

Naturalists looked even more closely at birds' role in social evolution. Many saw hints of a parallel evolution in which "inferior singers" in the avian universe like the bullfinch or nuthatch contrasted with the more varied and creative minds of the thrush or the blackbird, just as advanced races contrasted with primitive ones.[71] "Though the birds expressed themselves vocally ages before there were human ears to hear them," wrote Simeon Pease Cheney, "it is hardly to be supposed that their singing bore much resemblance to the bird music of to-day."[72] By 1919, social cultivation among birds meant that "the characteristic songs of the species are preserved, just as primitive human language passes from individual to individual within the tribe, and as the folk-songs of the various races of men have been handed down from generation to generation."[73] Like a crankshaft turning a pair of connected wheels, this *telos* of musical progression drove the evolutionary parallels between child and chick, bird and human that Spencer, Oldys, Sully, Weismann, and so many others believed in. It was a machinery binding humans to animals, and culture to biology, all through the power of beauty. "To those who like to think of the human race as closely bound to the rest of the animal world," wrote Sully, "it will be a very grateful thought that o' the pleasure which our ear drinks in from divine melody . . . even the tiny and fragile warbler of the woods has its own appropriate experience."[74]

## ANIMAL AESTHETICS

At the crux of these debates about birds and humans was the question of that tiny warbler's experience. "If our great biologist is correct," wrote science writer Grant Allen, "this theory of sexual selection thus becomes of the first importance for the aesthetic philosopher, because they are really the only solid evidence for the existence of a love for beauty in the infra-human world."[75] *Could* animals have an aesthetic experience? Did they have taste, sensibility, artistry? Did sexual selection mean that animals cared about what was beautiful? These questions guided many notions of animal behavior at the end of the nineteenth century toward a reductive view of animals as machines, with music situated in a uniquely human domain of moral development.

Aesthetic theory since the eighteenth century had constructed aesthetics to be deeply entangled in civil and moral development. Lord Kames reminded art critics in the 1760s, "We need only reflect, that delicacy of taste necessarily heightens our feeling of pain and pleasure; and of course our sympathy, which is the capital branch of every social passion."[76] Nineteenth-century concepts of aesthetics built on this theory that, to quote historian Marjorie Garson, "sensibility adopts and elaborates on the link between aesthetic responsiveness and social feeling."[77] Thinkers such as Hegel, Helmholtz, and Fechner speculated broadly about the ties binding together aesthetics, society, and morality.[78] Taste was the mark of moral cultivation and social bonding; according to historian John Finney, it was common knowledge that music "civilized, formed character and educated morally" those who learned it.[79] In Spencer's essay on music, these themes of moral and social development were very explicit. Music was "the basis of all the higher affections" because it had evolved to make humans sympathetic, "sharers in the joys and sorrows of others."[80] The ties between music, sociality, and moral sense were what made music "the difference between the cruelty of the barbarous and the humanity of the civilized."[81]

Evolutionists opposed to animal aesthetics imagined birdsong as a precursor to human aesthetics that lacked human discrimination. "Even if it be admitted that they [the birds] really appreciate singing," wrote music scholar and comparative psychologist Richard Wallaschek in 1893, "their discriminative taste for bird-minstrelsy could as little be called a feeling for music as their distinguishing one bird's plumage from another amounts to a feeling for painting."[82] "Is the bird's song a composition?" he had asked. "Certainly not . . . Birds have no conscious intention of charming."[83] Wallaschek's colleague Carl Stumpf similarly argued that birds lacked the capacity for taste; instead, he claimed that aesthetic ability could be measured by the human (rational) ability to recognize transposed melodies, the musical analogue of translating language. (Stumpf never learned that Pavlov's later work with dogs would disprove his claim.)[84]

Conwy Lloyd Morgan's landmark text *An Introduction to Comparative Psychology* offered a definitive description of this approach in 1894. Morgan's book outlined a law of behavioral study that has now served generations of scientists: "in no case is an animal activity to be interpreted as the outcome of the exercise of a higher psychical faculty, if it can be fairly interpreted as the outcome of the exercise of one which stands lower in the psychological scale."[85] "Morgan's canon" gave scientists a reductive approach to behavior that was intended to put to rest spurious claims about animal consciousness by excluding animals from the life of the mind. In Morgan's formulation, only humans had music; only humans had language; and, consequently, only humans had developed, rational souls.[86]

The reception of Morgan's work was positive, with reviewers calling it "rare" and "vigorous."[87] Later studies of animals often turned to it as the gold standard, and Morgan's canon became a widely accepted norm in laboratory psychology.

"So successful did Morgan's canon become," wrote science historian Gregory Radick in 2007, "that it now takes some effort to see it as anything other than the crystallization of scientific good sense."[88]

Morgan was particularly troubled by Darwinian sexual selection, in which animals expressed choice, emotion, and aesthetic sensibility through song. "Many biologists, for example, believe that birds select their mates from among numerous suitors because of their song or because of their bright plumage," he wrote. "Does not this, it may be asked, imply that she has a standard of excellence, and selects that mate which she perceives as the nearer of the two to such standard? But . . . it does not necessarily follow that she perceives the relation, or compares the two competing males to an ideal standard, or even the one with the other."[89] Morgan's explanation of a "standard of excellence" invoked Spencerian parallels between animals and human races, contrasting a "Somersetshire rustic's" bland reaction to Wells Cathedral (a response corresponding to the bird's lack of aesthetic reflection) to the "perceptive" mentality of aesthetic people (presumably of Morgan's breeding) to explain why birds and rustics lacked aesthetics.[90] According to Morgan's canon, one could not assume that either the bird or the human rustic had aesthetic ideals. After an extensive explanation arguing that birds did not make aesthetic choices because they could not be shown to make rational choices, he concluded that "we are bound by our canon of interpretation not to assume the higher faculty of interpretation"—aesthetics.[91]

Naturalists, especially bird lovers, fought this reductive vision of animal behavior by turning to music. Psychologist James Sully had already offered a reply to the Morgans of the world in 1879: "Mr. Darwin has recently taught us that certain birds display a very considerable amount of taste."[92] Anything less than this belief, later nature lovers argued, was an "egocentric standpoint" that privileged human culture.[93] Instead of the rat mazes or puzzle boxes that Morgan's followers used to measure animal intelligence in laboratory settings, a series of increasingly sophisticated studies of birdsong used music's listening skills to prove their point.

The first defenders of nature's music verged on the poetic in their enthusiasm for natural sound. Relying on prose, poetry, and musical scores, one author described the tiny sound of a flea's feet landing on a nightcap in 1841; decades later, in 1896, another wrote out the song of a faucet dripping in the key of B-flat major; and in 1910 the same tradition was alive and well in the sound of a moth tearing out of its cocoon in the night.[94] Comparable descriptions of "primitive" music were often limited to prose or simple transcriptions emphasizing the superiority of Western music.[95]

By the early twentieth century, however, studies of birdsong were becoming more sophisticated, both in their documentation of sound and in the use of music as evidence of animal aesthetics. Ferdinand Schuyler Mathews's *Field Book of Wild Birds and Their Music*, first published in 1904 and reissued in 1921 with new additions, documented the songs of over one hundred different species in detailed musical transcriptions. Mathews, a voice teacher, naturalist, and illustrator, used his musical expertise to author several sonic field guides in the early twentieth century.[96] Although he imagined the *Field Book of Wild Birds* as a bird guide for the general public, readers were asked to do some quite difficult musical tasks. One of his favorite birds, the hermit thrush, occupied more than ten pages of the book. Mathews argued that the hermit thrush understood basic harmony, could transpose his melodies, and—revealing a class of racialized musical categories that do not map neatly onto our twenty-first century imaginations—compared the bird favorably to Southern Negros, Scottish bagpipers, and Dvoràk.[97]

The culmination of Mathews's description of the hermit thrush was a comparison between the bird's song and the opening measures of the final movement of Beethoven's famous "Moonlight" Sonata. His example is eerily compelling, for the pattern of the thrush's song does sound rather like Beethoven's melody. But Mathews went much farther than noting their similarity. He argued that the bird's approach to harmony was the opposite of Beethoven's, claiming that although both examples built excitement through rushed phrases and harmonic shifts, Beethoven moved from dominant to tonic, while the hermit thrush took a more traditional route, from tonic to dominant.[98] Think for a moment about how many listening skills Mathews asked of his readers: identifying a wild bird by ear, listening carefully to it, identifying its musical scale by ear, recognizing the moment when the bird transposed its melody into a new key, and comparing the bird's approach to harmony with Beethoven's, which in turn demanded that the reader be familiar with the movements of the "Moonlight" Sonata.

Another bird widely acclaimed for its aesthetics was the American wood pewee. In 1904, Henry Oldys suggested that the wood pewee consciously grouped the phrases of its song in the form of a ballad, using the popular Stephen Foster song "Swanee River" as an example.[99] Oldys, a longtime member of the American Ornithologists' Union, had trained in law and worked as a government auditor before transferring to the Department of Agriculture, where he promoted game and conservation laws before turning his attention to studies of birdsong.[100] In one of his many comparisons between human and avian song, Oldys argued that in both the pewee's song and popular human ballads, melodies were repeated

FIGURE 1.3 Ferdinand Schuyler Mathews, comparison between the song of the hermit thrush (*top bird in illustration*) and the closing movement of Beethoven's "Moonlight" Sonata. Mathews, *Field Book of Wild Birds.*

according to an A B A B[1] pattern, suggesting that this technique made the wood pewee one of America's more advanced avian singers. In "Swanee River," the pattern follows each line of text; in the case of the wood pewee, the melody was different and moved much more quickly, but the pattern was the same: one phrase ("A"); a contrasting phrase that ended on a questioning high note ("B"); the first phrase again ("A"); and a final phrase based on the contrasting phrase, but with a lower-pitched closing note to round out the sequence ("B[1]") (see figure 1.4). Oldys seemed to take for granted that his listeners would hear a parallel in this comparison between the nonhuman musicality of the wood pewee and the lowbrow vernacular associations of "Swanee River" with minstrelsy and popular music, bringing the two styles into a kind of bridge between the higher stages of avian song and the lower stages of human music.

Amazingly, Oldys's argument was taken up and circulated in the ornithological community for forty years. Scientists such as Wallace Craig and Aretas Saunders continued to cite Oldys's analogy for decades as a counterexample to mechanistic explanations of animal behavior, and the wood pewee was often held up as a model of aesthetic capacity in the animal world by American ornithologists.[101]

FIGURE 1.4 Henry Oldys, comparison of the structure of Stephen Foster's popular song "Swanee River" with the song of the wood pewee. Oldys, "The Rhythmical Song of the Wood Pewee."

In 1944, Saunders reiterated, "all of the Stephen Foster melodies I know are built upon this plan"—by which he meant the wood pewee's plan.[102]

## AVIAN LISTENERS

The debate about animal aesthetics had surprisingly high stakes. Intellectuals of the late nineteenth and early twentieth centuries expressed tremendous interest in the music-making abilities of birds, mice, tigers, and other creatures. On the surface, these writers responded to Darwin's and Spencer's opposed theories of music's origins. But in practice, authors were often less interested in the details of evolutionary selective processes than in whether animals were capable of aesthetic creativity. The implications of animal aesthetics were huge, for music had a special place in social evolutionary theory at the borderline between human and animal nature. If animals made music, they were due the rights, status, and dignity that evolutionists like Spencer and Morgan attributed only to the most developed beings. Indeed, the holistic nature of social evolutionism meant that even one unexpected music-making creature might unravel all of evolutionism's social order, disrupting its tidy taxonomies of masculinity, breeding, and human uniqueness.

I conclude this chapter with a turn toward questions about whether birds were *listeners*, and the way those questions exemplified this disruptive potential. Music was the "language of the soul" for the nineteenth century. Nineteenth-century evolutionists found themselves asking if animals who understood this language had souls. In the natural history of the early twentieth century, these broad moralistic questions were succeeded by more detailed discussions about what it meant for a bird to listen. Songbirds in the writings of early twentieth-century naturalists are increasingly treated as beings with conscious feelings, goals, intentions, and self-reflection. In many ways, animal musicality had become about whether some birds were a new class of people, less privileged than idealized forms of European masculinity, but more respected than some of nature's other denizens.

Nineteenth-century debates about the place of songbirds in social evolution sometimes turned directly to the language of morality and the human soul. Writing in 1879, James Sully described the stakes of animal aesthetics as the stakes of moral status: "the new doctrine of Evolution . . . has naturally tended to raise the intellectual and moral status of animals by suggesting that in them are to be found the germs of mental qualities previously supposed to be man's exclusive possession. Among the attributes which science is thus attributing to the lower animals is the artistic impulse."[103] Darwin's opponents had often used the word "soul" to describe the moral status of aesthetics, rallying around a psychic dividing line between human and nonhuman musicality. German biologist August Weismann called the nonhuman animal "soul-deaf" in 1890, adding that "the same differences [between human and nonhuman] . . . must prevail in the different stages of the development of the human soul."[104] A medical doctor of the period used the same language to argue that the "material ear of the body" had to be distinguished from the "spiritual ear of the mind," the latter ensuring that "the intelligent comprehension of music, even by the higher animals, will always be more or less imperfect, because their soul is of a lower order."[105] Richard Wallaschek claimed, like Weismann, that it was "these peculiar qualities of 'soul' which have to be examined, and not a certain condition in the sense of hearing" in order to understand the shortcomings of birdsong.[106] Music scholar William Wallace even turned the accusation of soul-deafness against Darwin in 1908, calling him "psychically deaf" for not recognizing the spiritual properties that made human music unique.[107]

Those spiritual ears, thoughts, and emotions summed up in the word "soul" entered naturalists' writings in the early twentieth century through more focused questions about what it was like for birds to listen. Naturalists wondered what those nonhuman ears heard, and what beauty meant to them. "We must not forget," wrote Robert Moore in 1913, "that what is beautiful to our ears, may not be to a bird's," as he pondered the way the fox sparrow seemed to determine its own song only after carefully listening to those around it.[108] Several years later, another naturalist pointed out that the crow seemed to enjoy its harsh voice in the way an artist might. "Is he [the crow] not, in a limited way, a true artist, a composer as well as a performer? I ask it in all seriousness."[109] And in 1922, naturalist Richard Hunt wrote to *The Condor* to criticize Morgan's canon, arguing that the mockingbird could hardly be explained as a tiny machine, for he "not only takes a 'pleasurable satisfaction' in the results of his vocal efforts, but he does so because he *dwells upon* those results with pardonable satisfaction . . . I believe that the bird's interest in his own mimicry is 'artistic.'"[110]

Nowhere was the link between music and personhood more clear than in the contested sphere of the female's capacity to listen. Darwinian sexual selection and Spencerian social evolutionism located aesthetic capacity in the virility of men.[111] In the United States and Europe, theories of evolution that often represented women's primary role as reproductive coincided with polarized public debates about women's suffrage, rights, and roles outside the home.[112]

In a revealing episode, American ornithologist Chauncey Hawkins outraged his peers in 1918 with an article whose subtexts—published while Congress was attempting to ratify women's right to vote in the United States—reflected human politics a bit too clearly.[113] Claiming that song's primary function was to frighten rival males and break down the female's resistance to sex, Hawkins argued that sexual selection assumed an animal consciousness that could more easily be explained by the lively hormones of males. Males and females did not relate through musical aesthetics, he wrote, but through the brutal persistence and force that males needed to break down the "coy" reluctance of the female, a coyness he believed was nature's necessary counterbalance to the female's otherwise uncontrollable sexual impulses.[114] Song, he argued, wasn't about beauty or pleasure but about the fundamental differences between the sexes. "The male sings more vigorously because he is a male," he wrote, explaining that vigor was part of the male's necessary world.[115]

Although the article made some salient points about the shortcomings of Darwin's theory, its mechanistic approach and gender politics drew a rapid response. Aretas Saunders quickly wrote to *The Auk* to criticize Hawkins for his inability to tell calls from songs, as well as for Hawkins's claim that instances of female birdsong were aberrations caused by a hormonal imbalance, while Francis Allen mocked Hawkins for his naïve assumptions about "female constancy" and his inability to attribute agency to either gender.[116] "Shall we deny an equal appreciation of [song] to the female?" asked Allen, adding, "and if the female appreciates the beauty of the male's song, why should she not discriminate between the songs she hears and succumb most readily to the ardor of the finest singer?"[117]

Saunders, for his part, left his readers with a particularly striking description of courtship that represented male and female alike as sensitive and considerate listeners: "The Robin (*Planestictus migratorius*), in the late days of April when mating is in progress, may be found singing with its bill closed, the notes hardly audible for more than a hundred feet. At such times its mate is nearly always to be found in the same tree, evidently listening with pleasure to this whispered song, which is apparently sung for its benefit only."[118] It was a sound, he said, that he had heard frequently, and in both eastern and western robins; but neither in print nor in anecdote had he ever heard his experience corroborated, though birders today call this the "whisper song" and, just like Saunders, look for the listening female when they hear it.[119]

The listening robin, and her attentive mate, brought the questions of social evolutionism to a head. Were birds people? Was it foolish, or realistic, to imagine that a creature like a female robin could be such a discerning listener? Were artistic birds comparable to primitive humans, or to other kinds of people? Did naturalized notions of identity such as gender or savagery help scholars organize and interpret musical behavior within evolution's social order? Over the next several decades, such questions began to take shape as questions of practice rather than theory. The musical evolutionism that began as a print war developed into a viable profession in the late nineteenth and early twentieth centuries. For intellectuals who worked in this profession, differing perspectives on the status, rights, and dignity of nonhuman musicians and other "others" had important implications for the way music was heard and studied.

In the chapters to come, I examine the role of this professional practice inside the spheres of museum work, fieldwork, and laboratory research. Although the contents and form of questions about animal musicality shifted depending on the changing circumstances and context of each of these locations, the question of animal personhood remained a backdrop to the way studies of music were performed. In those studies, music became a tool that could reveal who was capable of listening, and, in many cases, who was capable of being heard.

# IDENTITY, DIFFERENCE, AND KNOWLEDGE

————

In the following three chapters, I examine the emergence of a professional study of songs founded on notions of animal life. Spanning the period between 1900 and 1945, chapters 2, 3, and 4 examine the classification, collection, and analysis of songs to show how music became a measure of difference in the twentieth century. The intertwined prehistories of comparative zoology and experimental physiology that I described in the introduction to this book serve as the basis for notions of identity and difference in professional music scholarship during this period. Building on that prehistory, these chapters link problematic constructions of animal identity to the creation of modern sonic knowledge. Identity, difference, and knowledge are the subject of this history.

# TWO

# Collecting Silence

## The Sonic Specimen

On the reverse side of this book's title page, below the publisher's name and thematic information, there is a string of letters and numbers. This code tells you how the book is catalogued in the Library of Congress. The world's largest library, the Library of Congress seeks to acquire and preserve a universal collection of human knowledge.[1] Since its inception in 1800, the library has acquired nearly forty million books, each organized according to topic. This book, for example, will be coded ML3900 if it is deemed to be about the social politics of music; HV4700 if it is a book about animal rights; and BD140 if it is a philosophical book about the origins of knowledge.[2]

But imagine that instead of sending the book to the Library of Congress, we wanted to store it in a natural history museum, sending it across the Mall from the Library of Congress to the National Museum of Natural History. There the book would become a natural history specimen, an object representative of natural knowledge. Instead of being organized by topic, it would be organized by its evolutionary relationships. As a biological object, the book is a derivative form of plant matter and would best be sent to the museum's Department of Botany. Its origin in North America suggests that it is probably made primarily of white pine pulp, *Pinus glauca*. The museum's larger specimens, which include things like pinecones, are stored in cabinets with pull-out drawers. We might, somewhat hesitantly, suggest that the curator place this book in such a drawer, somewhere between a branch of hemlock and a loblolly pinecone. We could then rather shamefacedly affix to it the label "Pinus glauca pulpus."

The imagined identification of a book in the botanical collection is not so different from the way specimens were represented when the National Museum of

Natural History opened in 1910. During the late nineteenth and early twentieth centuries, wealthy cities and individuals in the United States invested thousands of dollars to build lavish natural history museums decorated with crenelated turrets, columned facades, and delicate arcades in Washington, Chicago, Pittsburgh, and New York. Similar temples rose up in London, Vienna, and Paris, filling their halls with tens of thousands of specimens obtained on expeditions to Africa, Asia, and the Americas. In these places were assembled large collections that gathered together diverse objects: insects, plants, birds, shells, and fossils were joined in museum storerooms by anthropological collections that included tribal artifacts and the detritus of violated tombs. These collections were the basis of a science of identification. Specimens became a way to know who fit and who didn't, a map of nature that extended from animal skins to anthropological artifacts.

The music of humans and other animals had a place in these collections, with song collectors sending musical instruments, notated songs, and sound recordings to museums and other institutions of knowledge, where they became specimens within a sonic typology. Once institutionalized, songs became part of a broader natural typology, where they had to fit within the systematic habits of the library and the museum. Though new to most museums, musical recordings and instruments were welcomed into institutions. The majority of these musical artifacts were human songs, arranged to display racial and cultural development. Music historian Jann Pasler has shown, for example, how the Paris Conservatoire's musical instrument collection was organized to meet colonial narratives of racial development, while in New York, the musical instrument collections at the American Museum of Natural History and the Metropolitan Museum of Art were similarly crafted to reflect ideas about the relations between culture and natural history.[3] Occasionally the songs of animals, especially birds, were also arranged in collections; one archaeologist even suggested in 1914 that a museum of sound be built in the place of a traditional natural history museum, writing, "Bird songs are probably of as much interest to museum visitors as bird skins."[4]

In this chapter, I examine the inclusion of songs within institutional collections of natural knowledge. For the first half of the twentieth century, the evolutionary theories of development discussed in chapter 1 enabled broad comparisons between species and cultures. In the midst of these comparisons, the specimen took on a central role in the determination of biological identity. As naturalists and biologists developed new models for identifying species based on the visual inspection of specimens, experts in music adopted new models of sonic identification based on the examination of songs. I call these new forms of sonic information the sonic specimen, the musical analogue of the stuffed bird skins and preserved beetles arrayed in the natural history museum's specimen drawers. Like their biological counterparts, sonic specimens were compared in order to determine evolutionary relatedness, mapping development onto sound in the same way that naturalists mapped evolutionary change onto preserved animal bodies. As part of an institution of knowledge, song collections became elements of a broader narrative about evolutionary relationships between cultures, races, and species.[5]

Central to this practice was the challenge of understanding identity in terms of sound. The sonic specimen entailed the belief that sound, like its visual counter-part, could be used to determine identity. Identity and identification are recurring themes in this chapter, which explores the practice of aural identification within institutional contexts. What did identity mean to the museum curators, collectors, and librarians who imagined and built specimen collections in the early 1900s? How did those notions of identity relate to attempts to categorize music and music makers? How did attempts to order and organize musical identity reflect broader orderings of life?

In the pages that follow, I ask what we can learn by attending carefully to the institutional fate of avian and anthropological song collections as they became classified and categorized, particularly within the United States. During the late nineteenth and early twentieth centuries, insects, birds, shells, plants, and fossils were joined in museum storerooms by musical instruments, phonograph recordings, and collections of musical notation. Once they became part of an institutional collection, these sonic specimens were organized and classified in order to serve scholars as research objects, much as the books in a library serve readers. As researchers adopted this approach, the way comparisons were made between animal and human songs also took on new forms. Instead of comparing the songs of animals and human beings directly, musical "specimens" introduced a new comparison between the songs of humans and the bodies of animals. This was a meaningful change, for it meant that instead of comparing sound to sound, intellectuals began to weigh the value of different lives against the value of musical difference.

I begin my chapter by looking to the development of the "type specimen," a special kind of specimen designed in the early 1900s to serve as a new standard for the determination of biological identity. From that starting point, I turn to the fate of the wax cylinder recordings made by Jesse Walter Fewkes at the turn of the century. I use Fewkes's connections to the Harvard Museum of Comparative

Zoology and its director Alexander Agassiz as a point of departure for a broader examination of the translation of specimen-based identity to sound, where the singular type of biology had to be adapted to a more generalized notion of musical character in the examples I call "sonic specimens." Finally, I turn to the notion of the musical type, which drew on the biological precedent of the type specimen to suggest essential connections between species, race, and the classification of musical style. Within these interconnected histories, songs moved from museums to libraries and back as their meanings shifted from specimen to cultural artifact.

In the course of this chapter's investigation of the sonic specimen, I am also hoping to trace new contours on the surface of an otherwise familiar notion of identity. Today, identity plays an important role in many legislative and social constructions of difference through categories of race, nationality, gender, sexuality, and species. By turning to a history in which songs served as a means through which such categories could be discovered, I seek a meaningful context for considering the ways in which contemporary tropes of identity and identification are connected to much older attempts to evaluate difference, and to value different kinds of lives.

## IDENTITY AND IDENTIFICATION

Dictionaries of the early 1900s agreed that identity was a kind of sameness, the resemblance that made what could have been a misshapen pile of difference fit together as a single whole.[6] A year after the opening of the National Museum of Natural History, the 1911 *Concise Oxford Dictionary* defined identity as "absolute sameness; individuality, personality," and difference as "being different, dissimilarity, non-identity."[7] Identity was, speaking broadly, the likeness that made two things the same, and the differences that made likeness visible.

By the end of the 1800s, some biologists had adopted this marriage of sameness and difference as the definition of species. Biological species was nothing more than "an assemblage of individuals which agree with one another, and differ from the rest of the living world," as Darwin's defender Thomas Huxley put it.[8] But this was a very loose definition. How could a biologist tell whether a particular animal agreed with or differed from other similar-looking animals? And what, exactly, counted as an "individual" in the cloudy mist of the living world? One of Huxley's contemporaries explained with some exasperation that such broad definitions made it hard to avoid arbitrary categories: "For in every perception and judgment, and indeed in every sensation, the object reveals a twofold play of identity and difference. No two things are so much the same as to be indistinguishable in respect of *somewhat*, and that *somewhat*, even though it be only numerical, is a difference."[9]

In this amorphous garden of identity and difference, debates about biological identity flourished. Between 1900 and 1935, scores of articles were published that shared the opening title "The Identity of . . ." as they questioned existing species boundaries.[10] Species that were accepted in the sciences had, of course, a written history, often made up of descriptions dating back to the Renaissance and extending into the nineteenth century. These descriptions were considered an authoritative source, and were included in any article contesting the way a plant or animal was classified. Biologists began "Identity of" articles with textual exegesis, delving deeply and long into the textual history of a species. But when they applied these historical descriptions to actual examples, they often found that text alone did not tell them whether two living things were the same or not. Confronted by seeming discrepancies between authoritative texts and the creatures they purported to describe, biologists turned to a new authority: the physical body.

Historian of science Lorraine Daston situates natural history specimens in the field of botany at the center of this shift away from textual descriptions and toward a unique body called a "type specimen" or holotype. Daston shows how American botanists pushed to make the first recognized or discovered example of a species a new kind of specimen, the type specimen.[11] The type specimen or holotype was a single plant or animal body chosen to serve as the definitive reference for the identification of an entire species. Whenever possible, the type specimen would be the first known example of the new species, and the honor of naming it fell to its discoverer. The species' name was literally (with a label) and metaphorically (with nomenclature) attached to this individual body. Although some biologists believed that no single specimen should serve to resolve debates or questions about species identity, those who favored type specimens over texts dominated the debate in the United States and Europe by the early 1900s. With the adoption of the American Code of Botanical Nomenclature in 1907, the specimen became the central object of research in natural history, and the comparison of specimens was the single most important element in determinations of species identity. After 1907, the code was adapted for research in paleontology, zoology, and entomology, and other fields in the natural sciences. As one American entomologist wrote in the 1920s, the type specimen had become "the court of last resort in settling questions of identity."[12]

Type specimens brought together the visual display of the body, the twofold play of sameness and difference, and the systematic classification of typology. With the rise of the specimen came a shift in the way identity was understood, away from conceptual categories and toward specific, individual plant and animal bodies. Typology was no longer the ordering of ideas, but the ordering of specimens, with the type specimen serving as an exemplary case of the practice. "The specimen is, after all, the main thing," wrote one high school biology teacher in New York in 1907, the same year the Code of Botanical Nomenclature was adopted.[13]

With the adoption of the type specimen, the centrality of viewing pointed out thirty years ago by Donna Haraway and Mieke Bal became a critical element in the traditions that shaped not only biological specimens, but their sonic counterparts.[14] The visual tradition of specimen collecting contrasted with the challenges of quantifying and measuring sonic information, engaging attempts to apply this "court of identity" to sound in the differences between sound's mutability and the apparently stable nature of visual information. In reality, the specimen itself resolved a similar problem in biological science by substituting a preserved corpse for a living animal, exchanging information about behavior and its context for fixed information about morphology and anatomy. In the context of sound, the analogous solution was the inscription and transcription of sound in wax cylinders and on paper, making "living" sound a fixed object that could be measured. This solution, however, engaged scholars in two related practices that had considerable impact on the way sonic data was formulated: first, methods of recording and listening to sound were developed in response to criteria for data derived from the biological specimen tradition; and second, this tradition engaged scholars in a conflicted rhetorical opposition of life and death, sound and silence.

## SOUND IN THE MUSEUM

Moving from the National Museum of Natural History's botany collections to the shelves of artifacts in the Department of Anthropology, one finds a new set of bodies whose forms reveal identity and identification. Among the bowls and arrowheads are the ghosts of the wax cylinder recordings that once resided here, moved in the 1970s from natural history to the American Folklife Center, a division of the Library of Congress. The move was a retrograde of our imaginary deposit of *Pinus glauca pulpus*, in which sound left the museum for a library, instead of leaving a library for the museum. If we shift the clock into backward motion, the recordings tell one part of a much larger story about the institutional life of the sonic specimen.

The oldest of the museum's recordings was its collection of Hopi songs and speech on old Edison wax cylinders, collected in the 1890s during the Hemenway Southwestern Archeological Expedition. The cylinders had been inscribed on this hot and dusty expedition before being sent to the Smithsonian Institution's Bureau of American Ethnology in Washington, DC, in 1894. The bureau was devoted to pure anthropological research, with a sister institution focused on applied research in the Department of Anthropology at the National Museum of Natural History.[15] Though the cylinders probably served researchers in both divisions, in 1965 they were officially moved to the museum, as the bureau was incorporated into the department. There the bowls, arrowheads, and bracelets collected on the Hemenway Expedition still reside. But the cylinders are for the most part gone, except for a few copies: they moved again in the early 1970s, leaving the realm of natural history entirely for the Library of Congress, where they now reside in the American Folklife Center. Their life as specimens is (or, at least, is meant to be) over, for they serve a new tenure as part of the center's commitment to "creative expressions."[16]

In 1890, however, the cylinders were envisioned quite differently. Their collector, Jesse Walter Fewkes, wrote that year, "What specimens are to the naturalist in describing genera and species, or what sections are to the histologist in the study of cellular structure, the cylinders are to the student of language."[17] Fewkes was among the first to use the phonograph in ethnographic research, making the Hemenway cylinders a landmark in ethnology. Both complete specimens and the histological specimens prepared for microscope slides, called "sections," were familiar to him from his work as a zoologist curating the Harvard Museum of Comparative Zoology. There he had worked as Alexander Agassiz's assistant in the 1880s, overlapping slightly with Charles Otis Whitman and leaving just before Charles Davenport's arrival.[18] After a falling-out with Agassiz, Fewkes turned to ethnology and worked his way to the top of the Smithsonian Institution's Bureau of American Ethnology.[19]

Fewkes chose a fortuitous moment to champion the phonograph as the future of ethnography. He was a decade ahead of the collection of phonograph recordings begun by Carl Stumpf in Berlin in 1900, which became Europe's preeminent ethnographic sound collection under the aegis of the Berlin Phonogramm-Archiv (the subject of chapter 4). Over the next forty years, new song collections based on recorded sound included recordings of birdsong developed for research at Cornell University, the collection of non-Western music piled in the closets of the New School for Social Research by composer Henry Cowell, and the recordings of American folk songs made by the Lomax family in the late 1930s and 1940s.[20]

In a natural science dominated by the visual comparison of physical bodies, the phonograph provided a welcome "body" that, like a plant or animal, could be inspected with eyes as well as ears. In this context, recorded collections of sound can be profitably compared with photographs, for the two technologies had fundamentally different advantages from a research perspective. Photographs, unlike recordings, played a secondary, if valued, role in specimen collections. Museum dioramas of humans and animals served as models for photographs representing virtual images of nature that replaced uncaptured moments (and sometimes, as Aaron Glass points out, inspired the humans posing for diorama models to behave more "ethnically").[21] But from the perspective of dissection, morphology, and anatomy, replacing animal and cultural specimens with photographs meant a loss in information and the exchange of a primary source for a secondary one. Sound recording, in contrast, replaced in-the-field transcription with a perceived gain in information, exchanging secondary sources with something much more like a material artifact or "body."

Collectors hailed the phonograph as a tool that would make song collections "a branch of science," believing, like Fewkes, that this was music's answer to visual culture.[22] Historian Erica Brady has even suggested that the phonograph made possible the turn-of-the-century institutionalization of professional music ethnography.[23] Yet collectors were often stymied by the practical drawbacks of this new technology. Pitch and rhythm changed depending on the speed of the playback machine, making it hard to know what a song really sounded like in its original performance—worst of all was an irregular machine, which altered the pitch in unpredictable ways.[24] The British folk-song collector and composer Percy Grainger admitted that sometimes it was necessary to play a record hundreds of times to note down its details accurately, a process during which the record's fragile wax body would have deteriorated.[25]

In many ways, the ideal sonic body for ethnographers was the musical instrument, which had a literal body that could be treated like the body of a specimen. About a decade before Fewkes's phonographs found their way to the Smithsonian, many private collections of musical instruments migrated to museums, finding institutional homes in New York, Berlin, London, and Paris.[26] Their classification and display was grounded in the visual emphasis of the natural sciences, often resulting in large groups of beautiful or interesting instruments of dubious sound quality.[27] For many years the Crosby-Brown collection at the Metropolitan Museum in New York, for example, had several faux instruments that included a Mexican chocolate stirrer mistaken for a rattle.[28] Looking over this array of visual display, a curator in Paris wrote rather wryly that his collection of historical pianos and harpsichords reminded him of the taxidermical specimens of natural history, seeing "giraffe-pianos" stretching their frozen necks behind the glass wall as he passed their cases.[29] The author, Curt Sachs, had recently invented a classification system for musical instruments inspired in part by biological species classification, working with Erich Moritz von Hornbostel at the Berlin Phonogramm-Archiv.[30] Similar classification systems for instruments in France likewise relied on parallels between natural history and social history, organizing hierarchies between nations, colonies, and races through a parade of musical instruments.[31]

Neither wax cylinder recordings nor musical instruments were ideal sonic specimens. In lieu of better options, the object that was often used during the first half of the twentieth century for sonic display and comparison was musical notation. Before Fewkes's collection of Hopi songs, collectors had routinely relied on musical notation to gather together nature's music. Scores made good sonic bodies: they were flat, portable, and printable, and could be placed side by side for comparison. Unlike musical instruments, scores conveyed melodies; unlike the grooves of the wax cylinder, they made visual sense to a trained reader.

Early song collections in the nineteenth century were often made without reference to a sound recording, and were created by private individuals. The result was often unabashedly Westernized: collectors harmonized Native American songs like church hymns, and one guide to birdsong even introduced "natural" music with a hundred and seven notes of a dripping faucet in B-flat major.[32] By the twentieth century, similar collections were being published by institutions like the Bureau of American Ethnology, and often reflected institutional phonograph collections. The rhetoric of notation became a discourse preoccupied with science.[33] Ferdinand Schuyler Mathews explained that music notation was the best "scientific preservation" available for examples like his notation of a robin in four-part harmony (the result, with the robin singing along to a piano accompaniment, sounds very like a Bach chorale).[34] Privately, Mathews worried that his notations would be "hashed" in reviews by unsympathetic critics.[35] Another ornithologist working in the 1920s suggested that improved notation would require both alternate scales, like those in Gregorian chant or Chinese music, and a "battery of other instruments" beyond the piano to represent timbre, including the xylophone, banjo, zither, bassoon, and piccolo.[36] Still others hoped to replace or supplement musical notation with graphic scores and other symbolic images.[37] Even naturalists who hoped to replace musical notes with graphs or other images found musical references too useful to eschew entirely, often using images like a piano keyboard to map out the pitch and vocal register of various birds.[38]

Ethnographers, too, had energetic discussions about the musical bodies they constructed. In Berlin, Erich Moritz von Hornbostel and Otto Abraham argued in 1909 that scholars should systematize a series of symbols to adapt Western scores to non-Western sounds.[39] Many American collectors seem to have followed suit; Frances Densmore, for example, used many of the German notations in her transcriptions of Chippewa songs published by the Bureau of American Ethnography.[40]

One of the most intriguing collections was Benjamin Ives Gilman's transcriptions of Hopi songs, published in 1908 and based on Fewkes's recordings. Gilman's transcriptions created a hybrid between graphic and traditional notation, drawing the curving lines of performed slides atop fixed noteheads that represented exact pitches on an expanded staff. Fewkes, who had a tinny ear and struggled with music notation, had published extensively on the artifacts from the expedition but had done little with his phonograph records. He sought help from Gilman, a trained art historian with music literacy working with the Museum of Fine Arts, Boston. Gilman had an interest in music's evolution and agreed to transcribe the songs to create a more professional research object for the bureau. In a nod to the history of the American type specimen, Gilman called the collection his "hortus siccus," the insider's name for a botanical specimen collection.[41]

## AUDIOTYPING

As researchers struggled to find musical substitutes for physical bodies, they also faced the problem of developing a methodological analogue for natural history's systems of classification. Sonic specimens were adapted versions of biological type specimens, bodies whose interpretations were constantly redefined and debated as scholars struggled to accommodate the challenges of constructing a musical "body." How, then, did the biological identity made tangible in the type specimen translate into the musical identity of the sonic specimen? How should researchers categorize music? And how were their musical examples connected

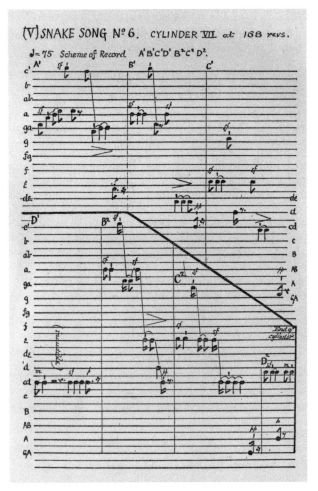

FIGURE 2.1 Benjamin Ives Gilman's "phonographic transcription" of Hopi song. Gilman, *Hopi Songs*, 100.

to biology's categories of race and species? Music scholars worried over these questions in the early decades of the twentieth century as they attempted to use song collections to classify musical difference. For answers, they turned to the same model that had allowed biologists to redefine the classification of species identity at the turn of the century: the type.

The backdrop for this discourse was an unstable history of biological classification marked by the transition from metaphorical typology in the eighteenth and nineteenth centuries to the twentieth-century type specimen. Specimen collections trace their origins to the age of seventeenth-century explorers and their ships' holds of exotica, when expeditions were motivated largely by the promise of selling valuable rarities. As the following century delineated the limits of the world, private and princely collectors built taxonomies of exotic objects in curio cabinets filled with large collections of preserved plants, animals, and insects. During the rise of the great colonial powers in the 1800s, these collections came to represent a country's imperial reach (sometimes in obvious ways: many nineteenth-century catalogues of Indian ornithological collections represent birds hunted by the British officers who authored the catalogue, but feature illustrations made by local Indian artists hired to do the time-consuming work of painting the figures).[42] Such collections offered research material for the growing interest in natural taxonomy that took definite shape during the late 1800s in an attempt to trace evolutionary relationships between species. By the early twentieth century, museums and individuals routinely arranged, displayed, and stored specimens in evolutionary ranks of species and subspecies, in drawers and "living" dioramas. By arranging species in order of relatedness, their viewers had immediate access to a visual taxonomy that educated and offered a highly organized frame for morphological research (study based on visual characteristics), making possible the comparison of large numbers of species.[43] For many twentieth-century biologists, the primary tool for scientific classification was the visual comparison and subsequent grouping and naming of different specimens.[44] The cornerstone of this project was, of course, the type specimen.

Like the textual descriptions that were superseded by type specimens, types also came from a literary tradition. Typology originated in biblical exegesis of the early 1600s, when theologians imagined patriarchs like Adam as "types" whose presence in the Old Testament prefigured the identity of Christ.[45] While explorers sailed to the world's far reaches in the seventeenth and eighteenth centuries collecting exotic curiosities, biblical typology was adapted to secular applications, and developed a broad reach in fields as far-ranging as biology, ethnography, music, and literary studies.[46] In the field of printing, the prophetic types of Moses and Ezekiel gave way to a different fixed form, the wood and metal block that held the form of a letter.

Typology entered natural history in the 1830s, when the famous naturalist Georges Cuvier adopted the notion of the type from the printing press to describe his new classification system, which would later be used as an adaptation of the Linnaean system.[47] Cuvier's types, unlike printers' blocks, were meant to arrange nature in a definite hierarchy in which lesser animals were represented by species that served as examplars (or types) of their particular genre. This was a typology of synecdoche, in which a strategically chosen part had the power to represent the greater whole. The many varieties of oblong fish who had hard scales, spiny gills, and toothy jaws were best represented by the perch, their type; while the leaf-cutter ant *Formica cephalotes* of South America, famous for its large colonies and social behavior, proudly stood as the type for the entire ant genus.[48] Historian Edward Eigen suggests that it was with rather coy self-awareness that Cuvier used different print types to show conceptual types, deploying various typefaces in *Le Règne Animal* to rank the categories in his prose from more to less general ideas.[49]

Over half a century later, when Cuvier's types gave way to type specimens, naturalists retained their mania for connecting words, bodies, and categories by synecdoche. They made up new words, like paratype, prototype, and holotype, for different kinds of specimens. One of the first articles in *The Auk*, the organ of the American Ornithologists' Union, suggested a long list of new words to be used in species nomenclature that included neologisms like "chironym" (for species names that had not yet been published) and "onymizer" (the person who successfully finds and names a species).[50] Several years later, paleontologist Charles Schuchert even suggested that biologists adopt words like "protolog" for the first written description of a specimen and "protograph" for the first known picture of a specimen.[51]

The bodies that these words represented reflected the tension between ideals and reality that was embedded in the notion of the type. Like Cuvier's types, specimens were supposed to represent an entire category of bodies. It was a heavy burden to place on the preserved remains of plants and animals, and naturalists carefully culled through their options to cultivate specimens that, they felt, met the ideal of a species. The modern museum preparator was expected to be a naturalist, sculptor, and artist, crafting what biology editor Frank Thone rhapsodically called "the exact form of the animal's lithe grace, its smooth waves of muscle, every vibrant detail that existed under its skin while it was still breathing and moving."[52] The story of the process through which such specimens were selected for display in the American Museum of Natural History's Hall of African Mammals, as told by Donna Haraway, shows just how important this interplay was between taxonomic practice and idealizations of the body.[53] Haraway's story demonstrates that the type specimen of a species was almost always male, and that display specimens were carefully curated to represent idealized postures and attributes ascribed to a given species.

This process of preparing and ordering specimens was just as important for those who collected and curated sonic bodies in the early twentieth century. But what was a musical "type"? Unlike an animal body, a song changed every time it was sung, and the recordings and transcriptions that served as sonic specimens were records of transient, singular experiences that had fundamental differences from the biological type specimen. A musical type was revealed by a whole collection of specimen melodies, not a single example—there was rarely a musical "type specimen." Instead, the cowboy "type" was found in the melodic pattern of American ballads for one scholar, while the English "type" was discovered by another in simple uses of church modes.[54] Naturalists, not to be outdone, compared hundreds of birdsongs to determine species types in the first half of the century. Aretas Saunders identified seven types in the song of the field sparrow (*Spizella pusilla*), and five for the song sparrow (*Melospiza melodia*).[55] Ornithologists at Cornell University managed to outfit a truck with film recording equipment in the 1930s, translating the songs of wild birds into the visible lines scrawled across the soundtrack, images they then compared under a microscope in their own search for avian song types.[56] The collector who financed that expedition, Albert Brand, sent some of the resulting recordings to a music ethnographer named Laura Craytor Boulton (one of the protagonists of chapter 3) for inspection.[57] Ethnographers working with the Bureau of American Ethnology on Native Americans worked especially hard to identify song types.[58] Before 1918, Frances Densmore plotted her collection of six hundred Sioux songs in tables and charts to determine five song types whose contours exemplified the Sioux tradition, while her colleague Helen Roberts borrowed Densmore's method to identify a type for the songs used in the Pawnee Skull Bundle Ceremony.[59]

Like biological types, song types navigated the disparity between ideals and thorny reality. Ideals in the case of music were closely tied to preexisting racial and national typologies. Music specialists often took for granted the "decisive leaning towards rhythm" of the French, the "innocent minds" of Italian melodists, and the "strenuous thought" of German musicians, hearing their own expectations in the songs of various nations.[60] British collectors Cecil Sharp and Charles Marson explained early in the century how a sufficiently large collection of songs could be used to scientifically demonstrate these associations, and eventually prove that "the scheme of a tune . . . develops racial traits [in time], just as the country has developed racial traits in the music of speech."[61] By the 1920s and 1930s, such empirical aspirations had reached into the laboratory, with experimental psychologists such as Carl Seashore suggesting that quantitative measurements of sonic specimens would "take us into the field of genetic studies of inheritance . . . and studies of racial types and the evolution of music in primitive peoples."[62]

Song typologies thus occupied the center of a three-part process in which songs were institutionalized and analyzed as objects of scientific research. First, collectors gathered them in the field. Collections were then assimilated into institutions and organized through typological practices. In the final and aspirational step of this process, song types were confirmed and analyzed in laboratory settings like Seashore's. Just as the lives of collectors shaped the songs they collected, the central step in this process of institutionalized classifications of music was also shaped by life's ambiguities.

The gap between real and ideal in the search for song types was particularly affected by problems of notation. Many of the specialists invested in song typing expressed concern that Westernized notations were making it difficult or impossible to study songs properly.[63] But those worries were often simultaneously framed by a desire that alternatives to Western scores reflect not a song's particular details, but its ideal typological form. As one collector of Russian folk songs explained, scholars needed to reduce songs "to their *simplest* form and, by omitting melodic ornamentation dependent upon the individual taste of the leader, preserve only the foundation of the melody."[64] It is worth reading closely the words of Helen Roberts, as she described the problems facing sonic typology in 1922 amid her research at the Bureau of American Ethnology:

> Not only is it difficult to seize upon and designate the peculiarities which distinguish certain types, upon hearing the song, but in reading over the notation they are not all equally clear to a musician, who must needs reduce the music to some simple formulae covering the structure, etc., in order to have them clearly in mind. Such a reduction is even more necessary for those whose unfamiliarity with the symbolism of printed music renders the subject still more complex. For the sake of obtaining the bald outline of the tune and the design which it formed, structurally, it seemed best for the time being in analyzing different songs to eliminate key signatures, musical notes, with their different values, all pitches less than whole step intervals, all measure bars and accents, all expression marks, in fact everything that might be considered to belong to the realm of color in music.[65]

Roberts suggests that although the collected song had to be accurately notated, its typological analysis was actually a generalization based on the removal of detail. Institutionalized studies of songs and their types in the laboratory are discussed in greater detail in chapter 4, but Roberts's words say much about the ways that visual representation, typology, and the sonic specimen were affected by the tension between real and ideal experiences of songs.

## IN THE FIELD

And what of the collector, the hunter in field and village who gathered songs alongside birds and bees? As I draw this chapter to a close, I move from the museum to the field, where the professional song collector provided a slightly different perspective on the work of institutional typology. In the museum, specimens were compared side by side in a space that stripped them of their original context. Typology, the final step in this process, allowed researchers to generalize about the characteristics that made up a body or artifact's identity by removing it from its original environment. This work of generalization was highly respected in studies of musical, as well as biological, difference. But in the field, identity was shaped by specific contexts, events, and relationships. Even scholars such as Roberts, who did both kinds of work, had to navigate the differences between locating specimens in the field and analyzing them in the museum. These differences were complicated by the fact that the work of collecting specimens was not as valued as the work of analyzing them. The specimens that were supposed to structure identity were themselves structured by the social hierarchies of science.

In the specimen's admission to institutions of knowledge, the work of collecting often came to seem less important and less prestigious than the comparisons that went on once animals and artifacts were in the hands of universities and natural history museums. Experiences of specimens in the field, be they animal or musical, were generally treated with less seriousness than their organization and identification within museum storerooms or laboratories. And curators and professors who worked with specimens in museums and libraries came to view collecting as clerical work, something that was done by junior scholars and, very often, by women.[66]

To some collectors, it seemed as if this mania for typology undervalued both the collector and the real, lived identities that he or she encountered in the field. Consider the following summary of one collector's five-page comparison of the institutional scientist and the field collector:

NATURALIST (institutional scientist)

Traces family, subfamily, genus, and species
Deals in Latin and Greek terms of resounding and disheartening combinations
Loses anatomy and markings in scientific jargon
Impales moths and dissects, magnifies, and locates brain, heart, and nerves
Neglects essential details and is not always rightly informed

NATURE LOVER (field collector)

Goes afield for rest and recreation
Appreciates the common things of life as they appeal to the senses
Identifies based on behavior and habits rather than species or anatomy
Does not care for Latin and Greek terms
Pronounces large silk moths to be exquisite creations

I compiled this resentful typology from the writings of Gene Stratton-Porter, a nature lover, lepidopterist, and novelist whose books include *Moths of the Limberlost* and *Music of the Wild*, the two books in which the Nature lover and the Naturalist were explained.[67] For Stratton-Porter, the rise of institutional collections marked the loss of natural knowledge based on living behaviors, valorizing instead the inspection of the dead. She used sound to drive home her point, explaining that only in the field would an experienced and dedicated collector hear the sound of a luna moth emerging from its cocoon. "There is a faint noise of tearing as the inner case is broken and the tough cocoon cut for emergence," explained Stratton-Porter, completing the sequence with silence: "Once in the air and light, if those exquisite wings make a sound it is too faint for mortal ears to hear."[68]

Like Stratton-Porter, the famed Hungarian song collector Béla Bartók was also a collector of moths. He was fond of insects, asking his son Peter to "send me at least a butterfly wing or beetle-thigh" when he traveled to Panama.[69] Insects found their way into Bartók's compositions as well, notably in "From the Diary of a Fly" in *Mikrokosmos* and the opening of "Musiques Nocturnes," the fourth movement of the *Out of Doors Suite* for solo piano.[70] But while Stratton-Porter lamented the literal deaths that had been institutionalized by the moth-impaling Naturalist, Bartók transferred this analogy of life and death to sound. A collector satisfied with mindlessly recording a song, he argued, would be like the entomologist or lepidopterist who would be satisfied with the assembly and preparation of the different species of insects or butterflies. If his satisfaction rests there, then his collection is an inanimate material. The genuine, scientific naturalist, therefore, not only collects and prepares but also studies and describes, as far as possible, the most hidden moments of animal life. . . . Similar reasons direct the folk music collector to investigate in detail the conditions surrounding the real life of the melodies.[71]

Music historians have pondered Bartók's representation of human bodies and races.[72] As a collector, however, Bartók imagined musical knowledge through the space between the animal body, the field collector, and the institutionalization of the specimen.

In the next chapter, I move from the institution of the sonic specimen to the experiences of collectors in the field. Stratton-Porter approached the differences between institutional research and fieldwork from the perspective of a deeply committed conservationist. She urged scientists to choose the path of the Nature lover over mass collecting, and argued that specimens should be replaced with photography. Bartók, whose career played out in both institutions and the field, believed that vast numbers of specimens (and musical deaths) were valuable, but also recognized that the experiences of living people would be affected if musical identity was purged of its context. The views of Stratton-Porter and Bartók, which offer a brief glimpse of a large and complex practice, suggest the importance of the perceived oppositions between institutional collections and specimens in the field.

At stake in those oppositions were two very different kinds of identity. In institutional comparisons of specimens, nomenclature and typology served to purge sounds of their particularities in order to classify and organize identity as an aggregate, a group or a class. Such categorical identities relied on notions of species identification to function within larger narratives of biocultural evolution, which told stories of large-scale changes that occurred through and across groups, rather than individuals. Objectivity, generalization, and the need for massive numbers of specimens attended work in the museum, library, or laboratory.

These notions of identity, however, failed to account for the role of the specimen in the field. There, identity was constituted through particular experiences and interactions between collectors and the sounds or animals they collected. Collectors and their subjects formed relationships that were both mercenary and affectionate. Animals and songs were hunted, prized, possessed, and imagined on a singular basis that seemed at odds with typological comparison, for the particularities that were purged in typological research were often intensely relevant in the field. Context, detail, and emotional relationships attended these particularities amid radical inequalities between collectors and the bodies and artifacts they hunted. In the next chapter, I leave behind institutional notions of identity in order to attend more closely to the experiences and voices of collectors, whose lives in the field held stories that institutional collections alone would never fully explain.

# THREE

# Collecting Songs

## Avian and African

The date is February 28, 1931. Somewhere in the heart of Angola, the search for biological identity is on a collision course with personal identity. Yesterday, a young woman discovered a new species of bird while out hunting specimens for the Carnegie Museum of Natural History with her husband. Together they shot two little warblers, a male and a female, who had been flitting four hundred feet apart on the slopes of Mount Moco. Today, as the woman packs away the birds' preserved skins, she is preoccupied with another collection, her growing pile of recordings of African tribal songs. As the woman travels south that day and for the rest of the trip, her songs will occupy more and more of her time. As her dual collections of animal specimens and songs grow, so does the need for a decision about which of her own identities will come into focus: woman, wife, scientist, explorer, musician, or scholar. Who is she? Who will she become? Who *can* she become?

———

Seven thousand miles away, on the eastern coast of the United States, another collision between personal identity and institutional identification is brewing. There, in a medical laboratory at Harvard University, a middle-aged man in shabby clothes catalogues library books. Ten years ago, he was a brilliant young zoology professor with high hopes and aspirations. He authored the first and only study of passenger pigeon calls, wrote groundbreaking articles about avian vocal behavior, and corresponded with Europe's leading animal behaviorists. Now, he is almost completely deaf, and his fortunes have shifted. With a string of part-time teaching jobs behind him, he does menial filing at Harvard with no clear plan for the future. Every day, he sees young Harvard students who will apply for the jobs he once hoped for, secure in their health, wealth, and friends. And who is he? Who will he become? Who *can* he become?

———

By the early twentieth century, the debates about musical identity that originated with Darwin and Spencer had become closely tied to notions of bodily difference rooted in biological specimens. But difference was just as present in the lives of those who collected and organized bodies and artifacts for the science of identity. For song collectors like the woman in Angola and the man at Harvard, difference was not a concept but a condition of life, the condition in which institutional science occurred. In that version of difference, animality was connected to lived experiences of gender, class, race, nationality, and other categories by material and discursive conditions. Entangled in those conditions, the work of biological and cultural identification was inseparable from the personal experiences of difference that determined who created musical knowledge, and how they did it. In this chapter, I turn from institutions to individuals, and to the ways their experiences became a mechanism of music's taxonomies.

This chapter focuses on the careers of two of America's first professional song collectors: music ethnographer Laura Craytor Boulton, and psychologist and zoologist Wallace Craig. My chosen protagonists are intellectual descendants of the zoology program at Harvard that produced Fewkes and the song collection at the Bureau of American Ethnology. Although other collectors such as Alan Lomax, Zora Neale Hurston, Olive Dame Campbell, and Fewkes himself are better known, the careers of Wallace Craig and Laura Boulton bring together the conditions that connected human social politics, the status of animals, and the construction of musical identity. Although they did not know one another, these two individuals shared ties to important movements in zoology and eugenics that had far-reaching effects not only on their work, but on sonic taxonomies more broadly. They occupied a middle ground in science's hierarchies, navigating flows of power that make their stories unique examples of scientific labor's reliance on the measures of difference it purported to discern.

Music scholar Suzanne Cusick once wrote that there is no place for "persons" within a historiography of large-scale musical epochs, styles, and epic changes, no room for individuals whose stories are rooted in historically and culturally specific, gendered, classed experiences.[1] This chapter is about specific persons and their experiences, about a kind of identity that is very different from biological or musical identity, and at the same time very closely related to its making. In this chapter I use names and stories to make space for the individual experiences that are unaccounted for by institutional history. Feminist historians like Cusick have sometimes used biographical writing not only to recuperate women's histories, but to endorse experiential values and alternative notions of selfhood.[2] As anthropologist Ruth Behar has pointed out, speculative biographies have served the needs of feminist and nonwhite scholars for many decades.[3] By structuring my contribution to that tradition through detailed descriptions of the flows of power that defined the practical work of song collecting, I ask my reader to imagine what it is like to be another person: a song collector, a colonial subject, or an animal.

The broader context of song collecting was one in which personal experiences of gender, race, nation, and species mattered acutely. Between 1900 and 1935, over one hundred collections of folk songs and exotic tunes, and over thirty gatherings of birdsong or insect chirps were published in English, French, and German.[4] Béla Bartók and Zoltan Kodály alone collected over ten thousand folk songs in Eastern Europe, while similar song collections were made of animal and "folk" cultures in the United States, Russia, Britain, Japan, Germany, and France.[5] These collections depended on long-standing relationships between Western governments and the resources and peoples within colonial economies. In the twentieth century, as song collecting shifted from amateur practice to full or part-time occupations, the economy of collecting remained strongly shaped by these economies of gender, race, and nation. Animals occupied a central role in the way that song collectors imagined and experienced these economies. Collectors called themselves "music hunters" or "song catchers," and spoke both of killing songs and of preserving them from endangerment and extinction.[6] In this world, how did songs-as-animals connect personal experiences of being different to the institutions and practices that positioned animals, women, nonwhites, and other "others" as placeholders in scientific taxonomies?

As I turn from institutional history to the lives of Wallace Craig and Laura Boulton, I am intervening in the historiography of epochs and epic changes to trace a different kind of connection between personal life and musical knowledge. Instead of exploring institutional notions of biological or musical identification, the stories I tell in this chapter help me identify *with* Wallace and Laura, a very different kind of identification. It is with that goal of identifying with an alternate self that I use "Wallace" and "Laura," while retaining the more formal convention of surnominal address for the other characters in my chapter. Being on a first-name basis with Wallace and Laura helps *me* identify with them—as an author gendered female, my sense of names is defined by the fact that my own name has never been "Mundy" among intimates. In asking my readers to be on a first-name basis with Wallace and Laura, I am asking for an act of imaginative identifying-with, instead of identifying-of. What would it be like to be one of these collectors? To be one of their subjects? To be one of the animals they stalked or hunted? How did those modes of existence become entangled with music's meaning?

The answers to these questions outline an economy of difference in which gender, species, ability, race, and other tropes of categorical difference were central to sonic knowledge. Contested measures of difference were at the heart of the project of song collection. Collectors faced widespread disagreements about definitions of biological identity; about bodily ideals; about the contested role of female scientists; and about an economy of knowledge dependent upon unequal privileges between bodies, spouses, species, and nations. In this world, the scientific evaluation of difference occurred within a broader context of real-life experiences of difference. Collectors were subject to the same process of comparison and evaluation as the objects they collected.

Difference, then, is the subject of this chapter, which turns to Wallace's and Laura's practices of collecting sound to explore how songs became a measure of difference in the twentieth century. I begin with Wallace's career, which emerged from the foundations laid by Agassiz at Harvard's program for comparative zoology in the 1880s. I then turn to Laura, whose career began, like Wallace's, with connections to Harvard. In the previous chapter, I argued that biological models of identity and identification were an important precedent for the classification and collection of songs. In this chapter, I hope to raise the possibility that identity itself is entangled in a more-than-human history of changing experiences of difference.

## ABILITY: WALLACE CRAIG AND
## *THE SONG OF THE WOOD PEWEE*, 1926–1943

I request that scientists and also non-scientific persons (if only they are accurate) will make records of the song for me. To make the record, you simply write the number of each phrase, 1, 2, or 3, as the case may be, while the bird is singing. This can be done by any person who is careful and reliable and has a sufficiently good ear to follow a tune. The only difficulty is that you need to rise and go out before daylight on one of the long days of early summer, in order to get a complete record. Few persons have the time to spare and the necessary interest and enthusiasm to do this. If you can take the time to make one complete record, it will be a valuable contribution. Please write to the undersigned for a copy of the directions for making a record:

Wallace Craig, Harvard Medical School, Boston, Mass.[7]

In 1926, a psychologist named Wallace Craig put this advertisement in *Science*, asking readers if they would take notes on the song of an American bird called the wood pewee.[8] If you answered the advertisement by writing to Harvard Medical School, you might have gotten a two-page letter explaining that the wood pewee sings three phrases. The letter would include musical notation of the three phrases, and describe them in prose in case you were not a musician. It would then ask you to keep track of the order in which your pewee sang its phrases using a number system of 1, 2, and 3 for the three phrases. As the ad in *Science* implies, you would also find out that the best way to do this was to get up at 3:00 in the morning and walk around a dark wooded area, trying to find a wood pewee before sunrise.[9]

When Wallace first published this request, his career was being dramatically reshaped by hearing problems that had begun to manifest in the 1910s. Born in 1876, Wallace had become recognized as an important if somewhat peripheral figure in the study of animal behavior by the time he died in 1954.[10] His work on the wood pewee was preceded by studies of the voices of various breeds of doves and pigeons, work that he conducted as a graduate student at the University of Chicago and continued as a young professor at the University of Maine until his hearing began to deteriorate. His work was characterized by a persistent faith in the ability of birds. "Each dove," he wrote in 1908, "is truly an individual, free to adapt itself to new conditions and thus to change its relations to society."[11] By the early 1920s, Wallace could no longer hear most bird songs, and he exchanged his original specialty in pigeon vocalizations for the song of the wood peewee, a simple, repetitive song that he believed could be transcribed successfully by volunteers.

The first half of this chapter is about the way that notions of ability and able-bodiedness defined the partnerships that Wallace cultivated in the economy of song collecting, many of them initiated by outreach like the advertisement *Sci-*

*ence*.[12] Aspects of these partnerships reflect the privileges of Wallace's position as an educated white man in the United States. But his privilege was premised on bodily ideals that he himself failed to meet as he became increasingly deaf. The story of his career traces his decline from privilege into disabling relationships as his body came to represent an imperfect specimen of manhood.

When Wallace began his graduate research in 1901 at the University of Chicago, those masculine ideals were tied to a distinct three-part career path that was expected of a young scientist. In the first step on that path, you would study and acquire a doctoral degree at a respected institution. From graduate work, you would have proceeded to your first job as a college professor. And from there, ambitious young men (for women had little to no access to this process) were expected to reach a third, aspirational rung, by making themselves the director of an independent research laboratory.

It was a path that all of Wallace's teachers had followed. They first witnessed it as students themselves in the 1880s, when their own teacher, Alexander Agassiz, charted this course. Agassiz, whom I discussed in the introduction to this book, was director of the Harvard Museum of Comparative Zoology. There he taught a cohort of successful students in the 1880s that included Charles Davenport and Charles Otis Whitman, the two professors who went on to make up Chicago's program in zoology while Wallace attended classes there. Son of the famous naturalist Louis Agassiz, Alexander Agassiz had considerable influence at Harvard, contributing over one and a half million dollars to the university and the museum during his lifetime.[13] In addition to Whitman and Davenport, Agassiz's students included the American ethnographer Jesse Walter Fewkes, who is discussed in chapter 2 for his work as the first song collector to use the phonograph on expeditions. In the 1880s, Agassiz began to envision a new stage to his career based on the success of independent laboratories in German institutions. Modeling his program at Harvard on those institutions, he decided to use the Harvard Museum primarily as an archive, moving his department's zoological research to an external laboratory. In 1887 he founded his own zoological station in Newport, Rhode Island, where students and assistants from the museum would come during the summers to conduct research. Whitman worked there from 1882 to 1883, and Davenport worked there from 1889 to 1892, while Fewkes assisted Agassiz at Newport from 1878 to 1886.[14]

When Whitman and Davenport left Harvard in 1892 and 1899 for jobs at the University of Chicago, they were both able to quickly advance to the third stage of this professional path. Whitman was made director of the prestigious Marine

| 1 | 2 | 3 |
|---|---|---|
| FROM DOCTORATE | TO PROFESSOR | TO LABORATORY |
| Alexander Agassiz<br>PhD, Harvard | professor,<br>Harvard University | Newport Zoological<br>Station, RI |
| Charles O. Whitman<br>PhD, Harvard | professor,<br>University of Chicago | Woods Hole Oceanographic<br>Institution, MA |
| Charles Davenport<br>PhD, Harvard | professor,<br>University of Chicago | Cold Spring Harbor, NY |
| Wallace Craig<br>PhD, Chicago | professor,<br>University of Maine | – – – |

FIGURE 3.1 The career trajectory from graduate student to laboratory director modeled by animal behaviorists and comparative psychologists, c. 1880–1940.

Biological Laboratory at Woods Hole, Massachusetts, and Davenport was made director of a school at Cold Spring Harbor, New York, which he transformed into a working laboratory in 1904 and, with the addition of the Eugenics Records Office in 1910, made into the nation's center for eugenics. (Years later, Davenport wrote a eugenicist obituary for Whitman that traced Whitman's interest in nature through photographs showing the physiognomy of his forbearers.)[15] The men would teach at Chicago in the winter and spring, and then spend the summers at their laboratories, assisted by select graduate students.

Wallace mastered the first two parts of this path successfully, gaining a PhD in zoology from the University of Chicago under Whitman and Davenport in 1908 and moving, after a brief interlude in North Dakota, to a tenure-track job at the University of Maine, where he began to assemble the means for a pigeon vocalization lab. Chicago, the first stage in Wallace's career, seemed to be an ideal choice for the ambitious young psychologist. Whitman and Davenport were respected scientists, and Wallace was particularly interested in studying Whitman's specialty, pigeon behavior.[16] At his home in Hyde Park, near the university's campus, Whitman maintained a large collection of living birds in a columbarium, which his students tended and used for their research. Because Wallace enjoyed music and played the violin and flute, Whitman suggested that he study vocal behavior, a topic that had been made compelling by the Darwin-Spencer debates.[17] Working with Whitman in Chicago, Wallace had access to his columbarium and was able to transcribe examples of his pigeons' vocal behaviors. He focused primarily on the ring dove, *Turtur risorius*, doing secondary research on the mourning doves and passenger pigeons in Whitman's aviary.[18] He spent his summers in Chicago or accompanied Whitman to the laboratory at Woods Hole on one of the strange treks in which Whitman had the bulk of the pigeon aviary shipped east with him on a train so his research was not interrupted.[19]

Wallace continued his work on pigeons when he moved to the University of Maine after his graduation in 1908, going through the motions of professional advancement that would lay the groundwork for the final move of his career toward an independent laboratory. He published the results of his pigeon research in a series of articles that sketched out a new model for thinking about animal behavior. Opposing Morgan's strict behaviorism, Wallace framed studies of vocalization as the key to understanding animals in the context of conscious social behaviors. Biological difference, in these studies, manifested through biological instincts that were integrated with a complex, learned social identity that Wallace imagined as animal culture. Darwin had argued in *The Expression of the Emotions* that animals' expressive trappings were often useless outside the context of mate selection, mere biological extravagances. Fifty years later, Wallace countered that emotional subjectivity not only helped creatures communicate with one another, but that emotional expression was a vital form of self-governance as well.[20] To Wallace, Darwin was being just as reductive as Morgan, ignoring the real meaning of emotional expressivity—of aesthetics—in animals. Wallace used words like "performance," "ceremony," and "convention" to describe pigeons with conscious social structures, behaviors he believed had to be studied with an anthropological outlook that assumed conscious intent.[21] And he compared the study of pigeons to the study of human cultures explicitly, arguing that bird ceremonies should be judged by the same criteria used to understand ceremonial behaviors in non-Western cultures. "Extravagance," he wrote, "does not prove that savage ceremonies are useless, no more does it prove that bird songs are useless."[22]

In the midst of publishing his findings, Wallace had to assume the role and trappings of a song collector in order to acquire data for his research, doing labor that he would have delegated to a graduate student at a larger university. He made many musical transcriptions of doves' songs and calls during his graduate research, publishing over seventy of them as musical examples in his articles before 1912. His most unique collection of bird calls was probably the small assemblage of passenger pigeon calls he published in 1911, three years before the bird became extinct.[23] Filed in *The Auk* between a pelican travelogue and a notification of the forest tern's rapid decline, the article contained over thirty transcriptions in musical notation of the rare bird's sounds. He hoped the article would contribute to awareness about the species and, possibly, to its rediscovery in the wild.[24]

As a collector, Wallace shared the language of hunting and preservation that defined specimens in the natural history museum. He had hunted his own specimens in the past, including a passenger pigeon he shot with a friend on the streets of Chicago in 1891.[25] Later in Maine, as his hearing declined and his situation became more desperate, his doves served a dual purpose as research subjects and food sources.[26] But Wallace also believed that vocal behaviors were a viable alternative to natural history specimens precisely because they allowed scientists to study living animals, instead of dead ones. He argued that musical examples contained the same level of detail about biological identity as natural history specimens did, claiming that vocalizations offered an alternative, behavior-based way to conceptualize species identity.[27] The passenger pigeon, which was nearly extinct by 1911, was a salient example of the value of vocal identification: "The voice has this further advantage as a mark of identification, that it cannot be produced in a dead bird, and thus forms an incentive to keep the bird alive."[28]

But Maine was a difficult place to make a start for a laboratory. The university was in an isolated spot, about ten miles north of Bangor in the small town of Orono. The main industry in the area was logging, and Wallace moved there just as the state's pine forests were dwindling, shipped south in the form of lumber and wooden storage boxes to Bangor.[29] Wallace found himself in this lonely place on a meager salary with no library and no funds for travel or research. He began to keep birds for study in his backyard, moving them in the winter to the second floor of his small house. He thought about studying hens or a larger variety of dove, because he and his wife could eat the eggs and "later we can put the specimens themselves in the pot."[30]

Then, in the early 1920s, Wallace began to have trouble lecturing because of his hearing problems. It is not clear whether he was fired or left his job, but he started writing to colleagues to ask for help in 1920, and left Maine in 1922 for a part-time position as a lecturer at Harvard. After teaching for two more years at Harvard, he was appointed as a librarian for one of Harvard's medical labs, where he did menial work indexing publications and writing abstracts throughout the 1930s.[31]

Harvard in the 1920s and '30s was a difficult place to be different. Women were confined to Radcliffe's grounds, and had to hire runners from Harvard to bring books from Harvard's all-male library to their classrooms and carrels.[32]

The young men studying in that library were overwhelmingly upper-crust, for the school attracted over seventy percent of the boys from Boston's wealthy elite families.[33] In the 1920s, the group was becoming more homogeneous, with the new president pushing for quotas to reduce the number of Jewish students by half by the end of the decade.[34] The teachers, too, were a tightly knit, homogeneous community known for hiring friends, family, and former students.[35]

Deafness was at best an awkward fit in this place of ambition, success, and youth. The deaf were often understood in the 1920s to be mentally defective; American law limited the number of deaf people who could immigrate, and many states prohibited the deaf from taking on certain jobs or from driving.[36] The clerical work Wallace did was a menial job normally reserved for women, premised on the assumption that he was unable to interact with students or faculty. Agassiz, the man who had taught Wallace's teachers Whitman and Davenport, had once even suggested replacing all of the male assistants at the Harvard Museum of Comparative Zoology with "cheaper assistants—Women" to do exactly this type of work.[37]

This unfavorable context was compounded by Wallace's social community, which was primarily made up of eugenics-oriented zoologists. Instead of becoming part of the growing advocacy for the rights and equality of the deaf, Wallace found himself asking for help from colleagues whose views of physical infirmity were closely tied to America's eugenics movement. At one point he asked the primatologist Robert Yerkes for help finding a job in the US War Department—Yerkes, who was a former Davenport student known for human intelligence tests and his work with Davenport's Eugenics Records Office.[38] He asked Davenport for help too, seeking a grant to complete his pigeon work at Cold Spring Harbor.[39] But Wallace probably knew or guessed that when Davenport founded the eugenics office at Cold Spring Harbor in 1910, he had argued that the deaf, the insane, the feeble-minded, and the criminal represented a "torrent of defective and degenerate protoplasm" that had to be removed from the nation's gene pool.[40] Davenport turned Wallace down, citing budget cuts.

Never one to sing his own praises, Wallace became modest to the point of self-sabotage. Years later he told one of his former Harvard students, "As soon as I came to Harvard I began to be deflated. I discovered that many of my students, even my undergraduate students, had better minds than my own. Since then I have given my research time to studies that are less ambitious and more exact."[41] It was from this position, in which physical disability meant legal, social, and emotional disablement, that Wallace issued his call for records of the wood pewee

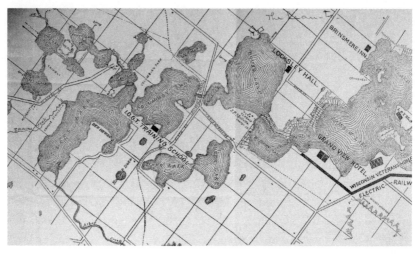

FIGURE 3.2 A topographic resort map repurposed by Mrs. Delbert G. Lean
in order to locate a Wisconsin wood pewee for Wallace Craig. Mrs. Delbert G. Lean
to Wallace Craig, Aug. 27, 1929, Folder 1, Correspondence Regarding
the Wood Pewee, New York State Archives, Box A 3157-55.

in 1926. He had a handful of his own transcriptions from the early days in Maine, but they were not enough for a serious study, and he could no longer hear the bird in the wild. Instead of giving up his research, he turned to song collecting, farming out the work of collecting birdsong to the able-eared. He received over 100 responses, and was able to compile over 100,000 examples of wood pewee phrases. The result, published in 1943, was the culmination of twenty years of observations sent in by friends and helpers, many of them aware of his deafness. It is also a deeply thoughtful study of the ways that music seems to make tangible the ability of beings who are radically different.

Wallace's advertisements in *Science* and *The Auk* yielded enthusiastic if some-times strange responses. He received poetic verses praising the bird's song, and letters lauding its voice. One observer sent him a musical transcription of a local wood pewee proudly marked "copyrighted" next to the date; Wallace returned the score after inking his numeric code in red over the notes.[42] Another respondent, wanting to make sure he knew exactly where her wood pewee was, sent him a detailed topographic map.[43]

Most of these letters, too poetic or too erratic, were unsuitable for Wallace's study. But Wallace did get about twenty respondents who consistently took

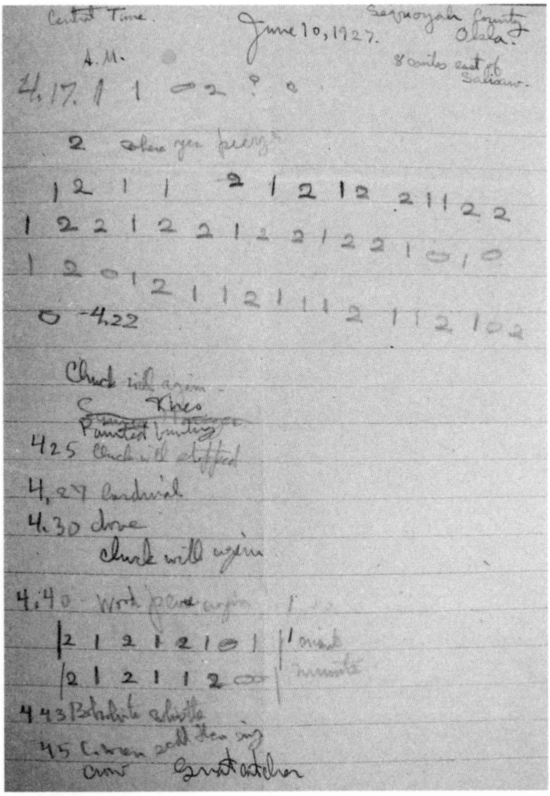

FIGURE 3.3 Notation of the wood pewee, Margaret Nice to Wallace Craig, June 10, 1927. Folder 1, Correspondence Regarding the Wood Pewee, New York State Archives, Box A 3157-55.

reliable notes and used his numeric shorthand. In a small way, these volunteers enabled him to retain his status as a researcher. Although he didn't have the institutional support that would normally provide for the livelihood of a laboratory director, his volunteers enabled him to take on the intellectual place of a lead researcher, albeit without a laboratory. He was also fortunate in being perhaps the only university-trained specialist in birdsong and comparative psychology in the 1920s and '30s in the United States. By the late 1930s, after he lost his institutional affiliation with Harvard, his unique niche became even more necessary, as it was more difficult for him to access recent literature and he had trouble staying current with the field. Instead of relying on university libraries and conferences, Wallace came to rely on friends, who did their best to keep him apprised of new developments through visits and letters.

Once Wallace had enough wood pewee records, he conducted a very close study of the song's phrasing, trying to tease out the relationship between innate behaviors and invented or learned ones. At a time when academic studies of animal behavior generally rejected nonhuman theories of mind, Wallace's ability to argue that birds understood musical concepts was quite unusual. His work on the wood pewee offered a comprehensive treatment of the birds' rhythms, melodic patterns, and phrase organization, and situated them in a nuanced and thoughtful analysis that acknowledged certain elements of innate biological instincts while championing the possibility of thoughtful, meaningful musical expressions by another species. It was a middle road that was not revisited by animal behaviorists until the 1950s.

I bring my discussion of Wallace to a close with one particularly curious example of his wood pewee work in which he addressed the problem of disability. In his 1943 monograph *The Song of the Wood Pewee*, Wallace discussed many examples of wood pewees who departed from the norm in musical behaviors that he had called in prior work "ceremonies." One of those departures was the case of a wood pewee with a speech impediment, a stutter or stammer.[44] The bird was observed by ornithologist Aretas Saunders, a naturalist considered by many to be the nation's leading expert on birdsong and one of Wallace's most valuable informants. It sang recognizable elements of the familiar 1-2-3 phrases, but combined them into much longer phrases, creating more complex-sounding songs than other birds'. Saunders wrote to Wallace that he believed the bird was a genius, a creative individual who fell outside the wood pewee norm.

Wallace agreed with Saunders that the bird was a genius; but he argued that it was not the bird's long phrases that made it so. Wallace noticed that the bird had trouble getting past its phrase 3, repeating it regularly. He wrote to Saunders that he believed the bird had a stammer, a known disability in songbirds that both Wallace and Saunders were familiar with. Considering the complex combinations that Saunders had sent him, he argued that the bird's real genius lay in its ability to make an advantage of its disability of vocal stammering. By embracing its need to repeat phrase 3, Wallace argued, the bird he called "Saunders' Genius" had made its limitations into an inspiration, creating a more beautiful song than those of its peers.[45]

*The Song of the Wood Pewee* was well received. Wallace and his former classmate Charles Adams, who had sponsored the publication through his position at the New York State Museum, received enthusiastic letters from scientists impressed and delighted with the work's meticulous collection of song, and its delightful presentation. The famous biologist Ernst Mayr even sent Adams a letter, saying the publication would be "*the* reference work of future workers in this field."[46]

Wallace's work on the wood pewee managed to substitute the work of an independent scholar for the institutional model of laboratory research popularized by his mentors Whitman and Davenport. It was a rare achievement, for by the 1940s, most song collectors required institutional sponsorship that enabled them to attach themselves to museum expeditions, laboratories, or institutional fieldwork. Operating outside of that institutional support, Wallace often fell behind professional trends. But he also retained an intellectual independence that startled and impressed his colleagues, who defended him fiercely. In 1945, the same former student to whom he had confessed his deflation at Harvard, Leonard Carmichael, shot down the idea of having students critique Wallace's work in order to keep him up to date. "If what Dr. Craig is writing is to have any lasting value," he wrote, "it must be because of his unique contribution."[47] Wallace, for his part, modestly hoped that his work was part of a greater whole, writing to Yerkes, "I should like to see studies made of the singing of all birds."[48]

## UNEQUAL PARTNERSHIPS: LAURA CRAYTOR BOULTON
## AND THE PULITZER EXPEDITION, 1930–1931

I turn now from Wallace's story to the early expeditions of Laura Craytor Boulton. For Wallace, the stigma of deafness preconditioned the work of hearing animals by placing him in relationships that were probably more emotionally and professionally disabling than his hearing was. Laura, too, experienced bodily difference as disabling, as a woman aspiring to a man's status in the sciences. Her story, however, is not one of decline, but of orchestrated success, for she leveraged her position as a white woman to create opportunities built on species and colonial hierarchies during her trips to Africa. Taken together, Wallace's and Laura's careers begin to tell the story of a broader relationship between personal and institutional knowledge, in which taxonomies of difference were not just the result of comparing and collecting specimens, but also the condition of possibility for those comparisons.

Unequal relationships were at the foundation of these expeditions. The bodies that determined identity, and the bodies that collected them, existed in a network of relationships through which hope, fear, violence, and love circulated. To better understand this process, I turn to Laura's experiences in Africa in and after the 1930s, a moment at the twilight of the lavish specimen-collecting expeditions sponsored by institutions and museums in the first half of the century. For Laura, these were also years in which the embodied partnerships of the collector were intensely real, years when she challenged her own identity as a woman, as a wife, and as a biologist doomed to the subordinate role of librarian and assistant. Her career is, in many ways, the inverse of Wallace's, for while he began as a professor and ended up a librarian, Laura started in the library and ended up a financially independent song collector.

Laura's path to song collecting began a decade before her first trips to Africa, when she worked as a librarian at the Cold Spring Harbor genetics laboratory on the northern shore of Long Island. Founded in 1904 by Charles Davenport, the lab was a center for genetics research and, after the addition of the Eugenics Records Office in 1910, the nation's center for eugenics. In 1919 Davenport became involved in a lengthy inquiry into the inheritance of musical talent, aided by psychologist Carl Seashore.[49] Much of the fieldwork for this project, which involved extensive interviews and heredity charts with targeted families, was done by unpaid women.[50] It was a year after this project began that Davenport hired Laura as a paid filing clerk for the Eugenics Records Office. Laura did well at Cold Spring Harbor. She had a background in musical performance and, given the timing of her employment, probably helped organize the vast family histories and charts dealing with musical inheritance sent in by Davenport's fieldworkers. In 1921, she coauthored an article with Davenport comparing racial traits like "suspiciousness" and "loyalty" in groups of schoolchildren categorized by ethnicity,[51] and within a year she was appointed the main librarian for Cold Spring Harbor's lab.[52]

Laura's role at the laboratory was that of a shadow-scientist: as a librarian at an important lab, she had better connections and resources than many aspiring young men in biology; but as a young woman in her twenties, she was also isolated from the experimental research that conferred influence and status. Dr. Davenport called her a "young biologist," but he treated her as a young girl, giving her rides in the country with the lab's other women and allowing her to take extra time off in the summers to pursue her other passion, singing.[53]

Instead of fading into clerical work, Laura managed to capitalize on the ambiguous status of collectors in the sciences. Women with science backgrounds were often considered good collectors, and a surprising number of them are responsible for the natural and cultural artifacts that populate American museums. Scientists' wives, such as Alice Eastwood and Florence Miriam Bailey, built many notable collections in botany and ornithology in the role of supportive spouse.[54] Alice Fletcher, Helen Roberts, and Frances Densmore were only a few of the women who, like Laura Boulton, turned to song collecting to develop an independent livelihood in ethnography. Though collecting specimens was vital for both natural history and ethnography, those who did it—men and women—were often seen as secondary in importance to experimental scientists. As historian Bruno Strasser put it, collecting was viewed as "mundane, clerical, or even trivial" by those in laboratories and universities.[55] Those who collected specimens were in a position of power in their relationships with plants, animals, and artifacts; but in their professional relationships, they were often on uncertain footing.

Laura began her first trip to Angola, the site of her first and only discovery of a new species, in this unenviably subordinate position. Rudyerd, her husband, was the staff ornithologist at the Carnegie Museum of Natural History in Pittsburgh. She had met him at Cold Spring Harbor, where Rudyerd began his career before a fellowship at the American Museum of Natural History in New York and, then, the Carnegie. With a handful of publications and a temporary position at a museum, Rudyerd had a strong foothold in ornithology, but he was unlikely to advance to the status of a laboratory director or university professor. Laura's position was more tenuous, though she was probably the more driven of the two. Her first expedition to Africa had been the year before, in 1929, when she had acted as Rudyerd's assistant. At her suggestion, a small portable phonograph was added to that first African expedition so she could provide an anthropological component to the trip's findings. But collecting songs took second place to her work for Rudyerd. As she explained in a press release on her return, "I was well armed with pitch pipes, reams of music paper, and a recording machine for the purpose of collecting Africa's music. Actually I spent most of my time shooting and stuffing birds, and injecting snakes and toads and frogs with formalin."[56] Added to this was the fact that Laura often found herself in the role of female companion for the upper-crust guests who joined the expedition. In 1930 on the trip to Angola, those guests were Ralph Pulitzer and his entourage: his wife Peggy and their son Seward, Peggy's French maid Georgette, and their butler, Georgette's husband Saunders.[57]

Wealthy patrons like the Pulitzers benefitted from sharing exotic holidays with a museum. They provided supplemental funds for the trip, while the museum provided experienced travelers, arranged the details of travel, and secured hunting licenses from foreign governments. African expeditions had included the likes of Teddy Roosevelt and the owners of the Macy's department store chain (including the apparently unstoppable widow Sarah Straus, who returned to Africa for her second trip in 1934 at age seventy-four). The 1930–1931 Pulitzer Angola Expedition was, for the Pulitzers, a kind of extended hunting safari. The Carnegie secured a license to collect specimens that included a number of the rare black sable antelope, which Ralph would be hunting. Ralph, in return, ensured that the trip was carried out in style.

The result was a dramatic mix of opposites. Rudyerd's stories about the trip composed a soundscape of "drums in the bush and Schubert on the record player" as he described the two hundred native porters carrying forty-pound supply boxes while the white explorers rolled along in a Ford touring car.[58] "The bush" was, in reality, layer upon layer of a changing society. The Portuguese colonial authority in Angola was at the top of this pyramid, and it enforced recognition of Portuguese settlers as the racial superiors of native peoples. But the Ovimbundu kingdoms in the central highlands and the Ovambo to the south, the two main destinations of the Pulitzer expedition, had remained independent from the Portuguese authority until 1904. They retained local traditions of regional royalty, village elders, and economies built on the vestiges of the previous century's slave trade, which had supported these pockets of independence since the seventeenth century.

Angolan porters were often socially mobile young men from these areas who would have joined trade caravans in previous decades to build wealth by selling local slaves for rubber on the coast, which could then be exported to Europe for a profit.[59] In the 1930s, porters were still the mechanism of trade in the region, positioning the Pulitzer expedition as a kind of alternative trade caravan. This colonial economy, however, was a fragile one. By 1961, the country broke out in a war for independence that soon turned to a violent civil war that persisted for thirty years. The people, plants, and animals of the region were devastated. Many of the animals the Boultons collected are now extinct or endangered, and many of the places they visited have been transformed into minefields.[60] The specimens gathered on the expedition, therefore, have increased in value, and are some of the only specimens of natural and cultural life in Angola before the war.

While the expedition's local members lived out Angola's complex divisions, the Western members experienced a kind of working holiday. The expedition began in London, where the group purchased supplies at the exclusive Fortnum and Mason, which were then shipped along with their human entourage on the German-run cruise line that ran along the west coast of Africa. Laura, thinking of the nighttime dances aboard that ship, later mused, "on all expeditions evening clothes were an essential equipment."[61] Once the party had landed in Angola, there were the expected potential dangers of wild cats and charging elephants. But the members also brought their own dangers from the West: one of the expedition's near-disasters came on Thanksgiving Day in 1930, when Ralph debuted a fancy cocktail shaker that manufactured its own ice when chemicals were combined in an internal compartment. The next morning, the expedition's violently ill guests realized that the chemicals in the base of the shaker had leaked into their drinks.[62] It was, quite possibly, the accident with the most potential for real disaster for the white leaders of the expedition.

Nonhuman animals were part of this peculiar economy. Africa's creatures have long been subject to the exotic fantasies of imperial eyes.[63] Animals' bodies were governed on expeditions through the hunting licenses secured by museums before the start of an expedition; animals were the currency of diplomacy on festive hunts arranged between colonial settlers and expedition members; and they were a temporary livelihood for the hunters and porters who were employed on the trip.[64] As museum expeditions traveled through a region, local residents would line up at the camp with an array of dead snakes, lizards, and birds in the hope of an easy day's wage.[65] As specimens, animals were at the center of the complex relations between Western travelers, colonial governments, and local populations in Africa.

In the midst of the hunting, preserving, and labeling of specimens, animals also became objects of genuine affection and curiosity on expeditions—in short, a few of them became pets. Laura's trip with the Pulitzers between 1930 and 1931 was filled with these pets. Photographs from the expedition show baby monkeys and wildcats, many in Laura's arms.[66] She cuddled for several days with two baby chimpanzees that the expedition had stashed in a railway company house at Huambo. Other favorites with the group included the affable and talkative African grey parrots and a family of civet kittens. By the end of the trip, a menagerie of over thirty animals accompanied the museum's representatives on the ship back to America, most of them bound for the Brookfield Zoo outside of Chicago. Laura gleefully noticed that her daily walks on deck with a chimpanzee were more popular with the other passengers than the cheetah brought on board by Woolworth Donahue.[67]

Laura, at least, believed that song collecting brought balance to this strange economy. "Bestowal of gifts establishes contact; a sharing of music inspires trust," she wrote in her 1969 memoir *The Music Hunter*.[68] She often saw herself as one of the girls, so to speak, in her travels. A musical encounter began with a classic exchange that Laura must have learned from other song hunters: first, she would sing a cheerful French folk song into her phonograph. Then she would demonstrate the playback mechanism before inviting her listener to play the same "game."[69] Later she carefully labeled the wax cylinder and entered the song in her expedition notebook.

Anthropology was a series of such sisterly games. Laura learned clapping games, drumming, and dancing from girls, basket weaving and the grinding of meal from older women, and a host of other womanly tasks. During her first trip to Angola, she learned about tribal adolescence rites from a teenaged girl named Iloy-Iloy, with whom she spent much of her time in Ovamboland. In anthropology's terms, Iloy-Iloy was Laura's informant, but Laura called her a friend. She described the relationship many times as a strong and mutual friendship, even comparing it to a romance.[70] Before the expedition left Ovamboland, Iloy-Iloy suggested the two become tribal sisters, and they exchanged bracelets in lieu of blood to formalize the ritual. The thick copper band is still on Laura's wrist in photographs taken decades later.[71]

Music brought other kinds of balance to Laura's life. As a biologist, she was ever in her husband's shadow. She often hunted alongside Rudyerd, but when the men wanted to go out together, it was Laura's task to stay behind in case someone needed to contact the camp. On one of the trips she did make with Rudyerd during the Pulitzer expedition, she accomplished what should have been the pinnacle of her career: two small birds, a male and a female, that proved on closer inspection to be an unknown species. She and Rudyerd skinned and stuffed the birds, and tagged them with the date, February 27, 1931. Laura and Rudyerd shared the credit for the discovery, and the new species was given a name in her honor: *Phylloscopus laurae*, Laura's Woodland Warbler.[72] When she and Rudyerd returned to the United States, the male specimen was sent to a special cabinet at the Carnegie Museum reserved for holotypes. The female, stored in a separate room, is his backup.

But in Laura's memories of Angola, the new species barely registered, and she sometimes forgot about it entirely as she became increasingly focused on music.[73] Birds began to migrate into anthropology. When she lectured for the staid National Geographic Society, she told of avian identities through loves, hates, and habits instead of bodies, asking why one bird fished with its wings spread, or why another wove intricate grass nests.[74] In Angola while the Pulitzers were hunting giant sable, Laura had learned about the dangerous onjimbi bird, whose call at night meant death to the Ovimbundu tribes; about the ungolombia, who complained to her mate that her fine garments would be soiled at the stream; and about the uugundu, whose ridiculous calls claimed that the crocodile, too, was a gigantic egg-laying bird.[75] Her favorite was the hornbill, whose mate locked her into a tree to raise the children:

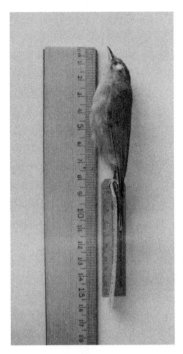

FIGURE 3.4 The male and female *phylloscopus laurae* that Laura and Rudyerd Boulton shot in 1931. *Left*, the male holotype or type specimen of the species; *right*, the female specimen, in top center of specimen drawer with topotype tag displayed. Author's photographs, courtesy of the Department of Birds, Carnegie Museum of Natural History.

A very interesting fellow is the ground hornbill or *epimumu*. He is to them the rain bird, and they have given the following words to his ventriloqual call:

Ngenda, ngenda, ngenda timbu lietu

To which the answer comes:

Kukande, kukande, ondomboyea tulimu.

The translation of the female's call is:

I am going, I am going away to my own village. Goodbye.

And the male replies:

You can't go. It is going to rain, and you must stay and hoe the corn.

His nesting habits would not appeal to the modern woman, I feel sure, because he seals his wife up in a hole in a tree and feeds her while she incubates her eggs and broods her young.[76]

Laura had no intention of being sealed up in a hole, and her sonic specimens were going to keep the door open. Unlike animal specimens, which had to be licensed through the Portuguese authority for the Carnegie Museum, music was unregulated and, for collectors like Laura, came with no apparent cost. So, too, were many of the photographs and films that Rudyerd took to document Laura's work as she sang and played games with the people in local villages. By the time Laura returned to Angola in 1947, she had leveraged her recordings and photographs into an independent career. She worked with a manager to develop a schedule of lectures for museums and clubs where she used slides, recordings, and musical instruments to talk about African culture and natural history. She registered for graduate school at the University of Chicago, placing herself on the first step in the ladder from doctorate to professorship to laboratory. She studied with anthropologist George Herzog, a student of Erich Moritz von Hornbostel, the Austrian music psychologist who had systematized song notation. In so doing, she insinuated herself into the lineage of those German laboratories that Agassiz had hoped to imitate decades before. She introduced herself to Hornbostel. She applied for a Guggenheim. And within a decade, she had also divorced both Rudyerd and biology, becoming a fully independent expert in her own field: a music hunter.

———

Wallace's and Laura's personal stories bring an important dimension to the meaning and measure of "difference" that institutional history does not convey. Difference was the foundation of their professional relationships, and those relationships were premised on bodily inequality. In their world, difference was not just the comparison of two bodies or songs that were otherwise autonomous and self-contained. It was a thing that shaped the flow of privilege and power through immediate, lived relationships between human actors, between humans and animals, and between sounds.

Why these two particular people, with their particular stories? Taken together, Wallace's and Laura's careers in the 1930s outline the connections between institutional ways of imagining musical difference in the twentieth century, and the personal histories of being different that underpin those intellectual traditions. Their successes and failures reveal a feedback loop, in which institutionalized inequality was the mechanism of the collection of sonic difference. Because the social and institutional networks where figures like Wallace and Laura were trained and professionalized—places like the Bureau of American Ethnology, Cold Spring Harbor, and the Carnegie Museum—were places that understood bodily difference through the lens of eugenics and comparative zoology, the bodies of researchers within that community were subject to its particular traditions of bodily comparison and assessment. Wallace's response to these circumstances, one fostered and enforced by his social world, was "deflation," a partially disabling emotional and social loss of confidence. Laura's response, in turn, was a shift from natural history to ethnography, made effective by her willingness to exploit her unequal relationships with animals and African singers. Though these responses differed, they indicate the degree to which unequal relationships—understood and justified through a particular American, scientific tradition of categorical difference—governed opportunities to define what musical difference was, and what it meant.

The effects of these conditions persist in the sonic taxonomies that are preserved in some of America's most prestigious museums and libraries. Wallace's wood pewee songs are in a carefully preserved box in the New York State Museum, which sponsored the publication of his monograph, a book that continues to garner interest today. Laura's birds, recordings, and musical instruments are stored in natural history museums and libraries across the United States. These collections represent the identities of the species and peoples with whom they are affiliated. But they also represent the personal identities of the collectors who produced them. In essence, they are products of the unequal relationships that made them possible, negotiations of gender, race, and species that shaped the basis of later knowledge about music and musical identity.

Music linked collectors to their informants, to animals, and to one another in an affective relationship that was deeply felt, and also constrained by hierarchy. For difference can inspire both contempt and passionate affinity, especially when we hear the radical difference of those unlike ourselves. "I am better, broader, wiser, happier for having heard the crickets and katydids," wrote one collector of insect songs in 1928; "for somehow there are points of kinship in our lives, even though our magnitudes and rôles of living seem so far apart."[77] For Wallace, such imaginative kinships prompted new ideas about the importance of song in the lives of birds. For Laura, they facilitated her faith in an economy of musical relationships that were, for her, also the means to professional advancement. Unprotected by law and imagined to be without cost, Laura's recordings from Africa supported her public lectures, gave her influence with universities and museums, and were eventually issued as commercial recordings by Smithsonian Folkways.[78]

Difference in their lives was both kinship and currency. For the song collector, difference was the negotiation of that kinship and currency, as it was conditioned by life's abilities and partnerships. Under the patronage of institutions, collectors' songs became part of a broader project to define and understand musical identity in a more-than-human world of sonic taxonomy. At the foundation of this project's many unequal relationships were the bodies and voices of animals. Animals' bodies lurked in Wallace's training in Chicago, his professorship in Maine, and his part-time work at Harvard. In Laura's career, birds' bodies were very literally the means and inspiration that enabled her to pursue song collecting in Africa. In the following chapter, I turn to the laboratory, where the stakes of animal life came to be very closely linked to music's economy of kinship and currency.

# FOUR

# Songs on the Dissecting Table

Dr. Hornbostel's problem was whether knowledge was worth killing for. The issue was made simpler by the fact that human lives were not required. Instead it was the life of a song he proposed to sacrifice on the laboratory's dissecting table: "When we laboriously dismember and disentangle the melodic thread with our scalpels—some fundamentally condemn such vivisection—we thus make the blood congeal; the living event must stiffen into an immobile corpse, at which it first becomes possible to identify the architecture of the now-intelligible whole."[1] With this gruesome assessment, Hornbostel proceeded to perform two vivisections that year, analyzing the Thai songs "Kham hom" and "Thai oi Kamen." He attached to this act the bloody and violent description above, disregarding the obvious fact that songs have no material anatomy, no legal rights, and no apparent need for protection.

Before judging the author of this deed as cruel or bizarre, take a moment to imagine him. In 1920 Erich Moritz von Hornbostel was the director of the Berlin Phonogramm-Archiv, a collection of sound recordings from around the world. In photographs his eyes seem mild, perhaps tired, as they peer out from a narrow pair of pince-nez perched over his carefully trimmed beard. Like Hornbostel, the songs in the archive were tidy and well groomed, each inscribed on a wax cylinder labeled with the date, title, and artists therein. In later years students remembered that these songs, Hornbostel's research subjects and the objects of his imagined scalpel, were also his *Herzenssache*, matters of the heart.[2]

The year 1920 was the start of a brief and tenuous decade of good fortune for Berlin, wedged between the shameful defeat of 1918 and the depression of the following decade. This was a city of learning and science, home to Alfred Einstein, Max Planck, Theodor Adorno, and Edmund Husserl. Berlin's scientists and scholars were especially proud of the city's legacy as Europe's center for experimental physiology. During the previous century, physiologists at the University of Berlin had made discoveries about virtually every organ in the human body, largely through vivisections performed on living animals.

The Berlin Phonogramm-Archiv took part in this tradition, acting as a small part of the university's new center for experimental research in psychology, the Berlin Psychological Institute.[3] The institute built on Berlin's scientific legacy to investigate new topics at the juncture of physiology and the far more ambiguous terrain of the human mind. Adjoining the Phonogramm-Archiv were rooms where psychologists studied vision, sight, smell, touch, and hearing. Their experiments relied on measuring instruments and techniques that had originated many years earlier in the bodies of animal subjects, applied now to the mind and not the body. When Hornbostel pulled a slender cylinder off its shelf, he faced a task of formulating empirical knowledge about music that his predecessors in Berlin had once confronted using animal subjects. The psychological laboratory presented Hornbostel with an ethics of knowledge grounded in notions of difference, including differences between humans and nonhumans, that allowed individual animals to be sacrificed for generalizable data. When music took the place of animals in Hornbostel's research, the ethical dilemmas of scientific knowledge-making transferred to the problem of generating musical data.

Data and its making are the subjects of this chapter. Hornbostel's description of musical vivisection brings together poles of knowledge and difference that framed the researcher's ethical dilemmas at the start of the twentieth century.[4] Out of these dilemmas emerged notions of musical knowledge-making as a practice demanding both scientific and moral discernment. In this chapter, I locate that discernment in the forgotten history of a machine, tracing Hornbostel's phonographs to a technological ancestry in the tools of vivisection, whose purpose was to extract knowledge at the price of nonhuman life. To look backward to that convergence of ethics and data is to adopt a dual fascination with knowledge and morality. What happens in the musical laboratory? Is anyone harmed? Does some kind of wrongdoing take place, or merely science? And what are we, turning from past to present, to do with our findings?

## INTELLIGENT REFLECTION

In the laboratory, there is a moment when emotions and intellect seem at odds. Your scalpel is poised over a perfectly healthy and undefended dog. You are about to cut a section of the tubing leading from her left kidney to her bladder, the left ureter, and then repair the tube by reattaching the two halves. The dog will die. But your alternative is to practice this difficult procedure for the first time on a human patient, knowing you will almost certainly fail to attach the tubing properly in your initial attempt. You set your teeth and the scalpel descends.

This dilemma was posed in 1919 by Dr. Thomas S. Cullen before a subcommittee of the United States Senate charged with investigating the moral implications of the vivisection of dogs.[5] Vivisection in the late nineteenth and early twentieth centuries was a controversial but widespread practice in Europe and the United States. Doctors and biologists like Cullen had been cutting into corpses for centuries to better understand anatomy. But in the mid-1800s, European physiologists began to experiment on living animals to uncover the functional workings of the body's organs. Their discoveries were remarkable, locating the inner mechanisms of the ear, the heart, the lungs, and the nervous system. But they were also gruesome and painful, and by the turn of the century, anti-vivisection societies had sprung up in London, New York, and Paris. Vivisection and dissection came to be seen as a significant moral quandary that affected not only scientists, but intellectuals of every kind.

Dr. Cullen's dog was not, by the 1920s, the only nonhuman to whom these ethical concerns applied. Literary critics deplored the "criminal vivisection" of sonnets while musicians worried over the vivisection of sonatas.[6] Even birdsong could be dissected, as Wallace Craig intimated in his comparison between sound and visible organs.[7] Art objects threatened (or honored) with vivisection and dissection included stories, poems, opera libretti, musical instruments, and songs. The analysis of music seemed particularly prone to this animal morphizing. Here are a few of the arguments for and against musical vivisection between 1920 and 1933:

> The protest is often heard that cruel vivisection is being practiced; but it is only by some such laboratory method as this, that elements [of music] can be isolated and studied. (Karl Eschman, 1921)[8]

> Vivisection may be necessary for medical progress, but vivisection of [musical] art for the better enjoyment of art is abhorrent. (Oscar Sonneck, 1928)[9]

> The analyst may dissect the body of music, uncover its structure, exhibit the interrelationship of phrases and sections, yet fail to find the soul that animates the form. (William Fisher, 1929)[10]

"Vivisection is only occasionally and incidentally the infliction of pain, and anti-vivisection is not really a campaign against pain at all. The real campaign is against the thrusting of a scientific probe into mysteries and hidden things which it is felt should either be approached in a state of awe, tenderness, excitement or passion, or else avoided" (H. G. Wells).

Without taking sides about this statement . . . it is possible to apply the principles which are implied to the creative impulse in contemporary music.[11] (Basil Maine, 1932)

Nothing can be disgusting if we approach it in a scientific spirit, and the dissection of romantic emotion [in music] may teach us much about the psychology of musical expression. (Edward Dent, 1932)[12]

To dissect the experiences that are aroused by a single chord or phrase . . . I shall adduce a small quantity of strictly objective experimental research, carried out by other psychologists, which is based chiefly upon measurements of the breathing, the blood circulation and the muscular contractions that take place during listening to music. (Philip E. Vernon, 1933)[13]

I would like to take a moment to explore the mindset of the actual vivisectors and their opponents who inspired these debates about cutting into things literary, musical, and otherwise. Opponents of vivisection were particularly visible in Britain, Italy, and the United States at the end of the nineteenth century, campaigning to have the practice regulated or outlawed. They included doctors, intellectuals, and, interestingly, advocates of women's rights, who compared women's plight to that of animals and found in anti-vivisection societies opportunities for leadership.[14] Their objections echoed those quoted earlier: that vivisection was cruel and inhumane, and that it destroyed the thing it purported to study.

For the vivisector, the crux of the vivisection debate was a false conflict between intellectual and emotional values. Advocates of vivisection believed that scientists had to set aside personal emotions in the pursuit of knowledge. They argued, as Dr. Cullen did, that once emotional prejudice was removed, it was clear that the benefits to humans of animal research outweighed the suffering of a few animals—indeed, witnesses at the US Senate hearing even testified that altruistic dogs would approve of canine vivisection, because it also helped veterinarians treat dog-kind.[15] One American zoologist went so far as to claim that vivisectors were more reverent than other men, because the first vivisection occurred when God anesthetized Adam and created woman from his rib.[16]

This "scientific spirit" of emotional distance was considered a necessary part of the scientist's ethical life. Many years before the Senate hearing on canine vivisection, the *North Carolina Medical Journal* reported approvingly of the role vivisection played in German medical training: "Every medical student has a struggle with his instinctive feelings in making acquaintance first of dissecting-rooms and museums of morbid anatomy, then of hospitals, and especially of surgical cliniques; but every one knows these feelings must be repressed, or he will never have the steady hand and eye which give poise and self-control in a critical emergency."[17] The ethical argument was straightforward: because the sacrifice of animals and emotions saved lives, overcoming one's revulsion at the destruction of the individual was a critical part of the scientist's moral training.

In Germany, this attitude was more widely adopted than in Britain, Italy, France, or the United States. Ernst von Weber's anti-vivisectionist tract *Torture Chambers of Science* (*Die Folterkammern der Wissenschaft*) launched a brief public debate in German-speaking lands in 1879, but never led to widespread legislation of any kind. Instead, most Germans adopted the idea of "intelligent reflection," morality guided by reason and influenced by respect for science. Rudolf Virchow, a vivisector selected to lead a state commission considering anti-vivisection arguments in 1880, spoke of this "verständigen Betrachtung" as the antidote to the anti-vivisector's emotionally motivated morals, which he condemned as thoughtless and impulsive.[18] The vivisection of music, animals, and other things was thus viewed as a sign of intelligent morals or arrogant brutality, depending on which side of the debate one took. By the early twentieth century, vivisection debates in Germany had almost completely disappeared, and the public seemed indifferent to the use of animals in research.[19]

## THE MUSIC LABORATORY

In 1920 the Berlin Phonogramm-Archiv was at its peak as part of an institution of German science. A decade earlier, the Berlin Psychological Institute had finally moved out of the three cramped rooms at number 5 Dorotheenstraße, where it had been squeezed between classrooms in theology and German. Professor of psychology Carl Stumpf had inherited the woefully underfunded experimental psychology program at the University of Berlin from his predecessor, Hermann Ebbinghaus. Thanks to Stumpf's campaigning, Ebbinghaus's tiny psychological laboratory was transformed into a well-lit, ten-room research institute on the top floor of a new building at 95/96 Dorotheenstraße, and later moved to even larger rooms in the next block. The institute's new home included a library, laboratory space, a darkroom for optical experiments, and storage space for Stumpf's collection of over three hundred acoustic and optical instruments. It also included a special room set aside for the Phonogramm-Archiv (Figure 4.1). The archive was something halfway between a museum collection and a laboratory experiment, a place to study aesthetic psychology in empirical terms.[20]

Empirical science was very much part of Berlin's culture. In the 1840s, professor of physiology Johannes Müller had pioneered a new experimental approach to studies of the body at the university, using physics and chemistry to measure subtle physiological effects. This was where vivisection was at its most useful, for Müller and his students could attach mechanical instruments to the nerves, eyes, and ears of living animals to record and measure tiny motions. Müller's students were hired in laboratories across Germany, where they became famous for studying the physiology of the senses, a field Müller's student Wilhelm Wundt dubbed "experimental psychology."

Wundt's laboratory in Leipzig was Stumpf's main competition. Wundt had built an impeccable reputation using laboratory experiments with human and animal subjects to measure muscular sensations, nerve functions, optical illusions, and a host of other sensory effects. In his later career, Wundt turned to the topic of *Volkerpsychologie* or cultural psychology, a process he argued could not be measured experimentally in the laboratory.

It was around this time that the University of Berlin began to look for a psychologist who could develop its small psychology laboratory into an international powerhouse capable of competing with Wundt in Leipzig. Historian of science Mitchell Ash has traced the institutional motives behind the university's choice of Stumpf, a widely respected psychologist and intellectual who took up the post rather reluctantly.[21] Stumpf questioned the value of Wundt's empiricism from the beginning, and urged the Berlin program to move in a more philosophical direction.[22] In his introductory seminar, Stumpf placed equal emphasis on abstract reflection and experimental observations, training his students to trust their introspective abilities as well as their experimental results.[23] The popular class cemented Stumpf's authority in the eyes of the university, and he was able to leverage strong funding for the institute. By 1911 Berlin had the best-funded psychological laboratory in Germany, possibly the best in the world, with a budget of 3,400 marks—Wundt's lab in Leipzig had 2,000 marks, and the smaller labs in Göttingen and Würzburg had less than half as much.[24]

Experiments at the Berlin Psychological Institute reflected Stumpf's penchant

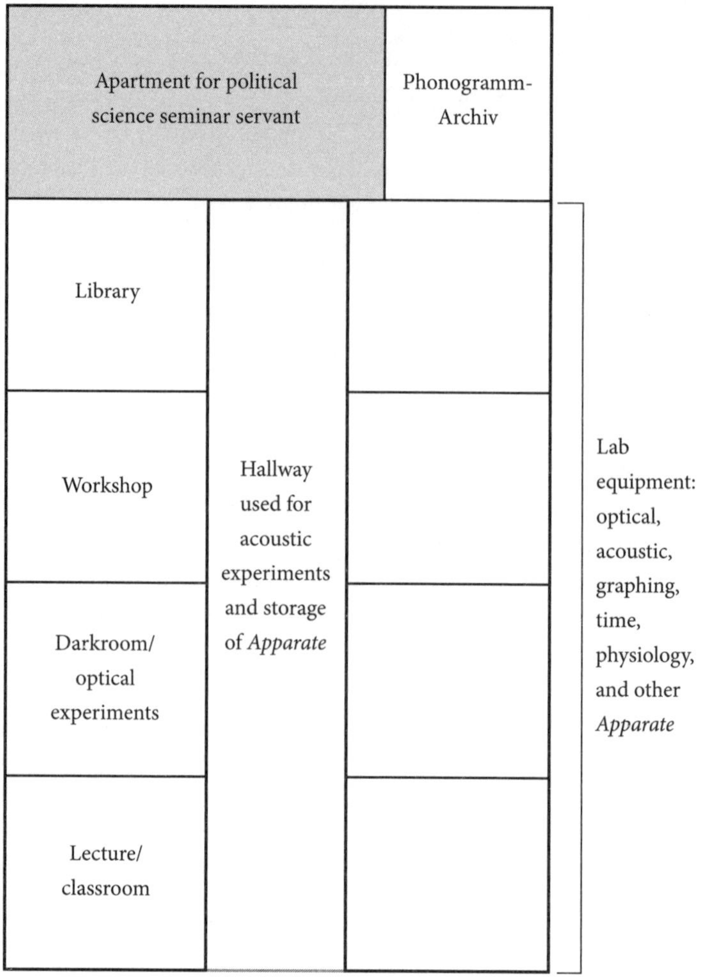

| Apartment for political science seminar servant | | Phonogramm-Archiv |
|---|---|---|
| Library | Hallway used for acoustic experiments and storage of *Apparate* | Lab equipment: optical, acoustic, graphing, time, physiology, and other *Apparate* |
| Workshop | | |
| Darkroom/ optical experiments | | |
| Lecture/ classroom | | |

FIGURE 4.1 Approximate layout of the Berlin Psychological Institute, circa 1905, as described by Carl Stumpf in "Das Psychologische Institut."

for the abstract and philosophical. With Otto Pfungst, he led a commission investigating the psychology of Clever Hans, a horse who appeared able to perform complex mathematics (but, Pfungst showed, was actually responding to subtle signs of pleasure from his owner). Stumpf's students Max Wertheimer, Kurt Koffka, and Wolfgang Köhler conducted extensive studies of optical illusions and acoustic perceptions, eventually founding the new school of Gestalt psychology. Yet for Stumpf, the brass ring of psychological research was not Gestalt but the marriage of philosophical introspection and experimental testing that Wundt had found impossible: the empirical study of culture.

With the Phonogramm-Archiv, Stumpf sought to reach for this brass ring. Today the archive is a kind of art library, its songs stored in the Berlin Ethnological Museum and selectively sold in mp3 format. But originally it seems to have been Stumpf's answer to Wundt's *Volkerpsychologie* and the Leipzig model of empiricism. In the 1890s, Stumpf had engaged in a heated debate with Wundt over the value of a musically trained observer in the laboratory, a print war that historian Alexandra Hui suggests was fueled by the diminishing value of musical expertise in experimental psychology.[25] A successful song laboratory directed by a musical expert would have refuted Wundt's disparaging assessments of musical expertise and cultural empiricism. The purpose of the Phonogramm-Archiv, Stumpf wrote in 1911, was hardly the "tomfoolery" some might imagine it to be, but an attempt "to return to ethnological studies and connect them to experimentation."[26] Such a connection would indeed place the institute at the forefront of psychological research.

To accomplish this goal, Stumpf needed a researcher with a strong background in experimental science and extensive musical training. He chose Hornbostel, who had arrived at the institute in 1905 to pursue postgraduate work in psychology, physiology, and anthropology. Hornbostel had grown up with music, listening to Johannes Brahms and Artur Schnabel play in his mother's living room as a child.[27] He played the piano, composed, and could transcribe by ear. But he also had unshakeable scientific credentials, with an undergraduate degree in physics and a doctorate in chemistry from the University of Vienna. By the 1920s, Hornbostel had published in respected science journals on optical and acoustic psychology, and had sent groundbreaking studies of non-Western music into print.

For the Phonogramm-Archiv to work as a laboratory, Hornbostel had to rely on his scientific background in order to create convincing experimental evidence. Psychoacoustic experiments offered some precedents for the measurement of sound and its perception. And physiological research offered the gold standard of laboratory evidence through the techniques of vivisection.

Though vivisection was not part of Hornbostel's work at the Berlin Psychological Institute, it was part of the intellectual vocabulary he brought to his job. Fifteen years before recommending his Thai songs for vivisection, he told the International Society for Musicology why it made sense to take the song of a Hottentot and start "dissecting it with a tonometer and a metronome," playing on the famous Parisian dissection of the "Hottentot Venus" Saartjie Baartman.[28] Demonstrations using dissected and vivisected animals were a routine part of the seminars in physiology attended by Hornbostel and other members of the institute. After his student days, Hornbostel continued to read current literature on animal research. He kept careful notes on Catherine Corbeille's study of sound's effect on decerebrated rabbits; Yoshitsune Wada's experiments on the tympanic membranes of canaries and turtles; and Jellinek Auguste's study of doves with partially removed eardrums. He read T. B. Manning on the physiology of hearing in goldfish, Käthe Berger on hearing in reptiles, and Harry Allard on musical technique in katydids.[29] Hornbostel even produced his own study on the relevance of birdsong to the psychology of music.[30] Building on these foundations laid by physiology, the Phonogramm-Archiv could develop its own standards of musical knowledge.

## DATA

"Data, data, and more data—by the hundreds of thousands!" Hungarian folk-song collector Béla Bartók once cried in his enthusiasm for the wax cylinder.[31] For early music ethnographers, these recordings were the salvation of musical data. Historian of science Alexandra Hui has compared the wax cylinders of the Phonogramm-Archiv to the microscope.[32] But they were also a musical body, the sonic analogue of the bodies in scientific specimen collections.

It is difficult to know exactly what Hornbostel meant when he ascribed mortality to these musical specimens. To study the bodies of wax cylinders does actually kill them, albeit slowly. The wax surface that holds the groove in which sound is encrypted wears away slightly with each use. Over time, the sound quality degrades until eventually it is lost entirely. It takes many repetitious hearings to study a song, and it is true that "Kham hom" and "Thai oi Kamen," the songs Hornbostel studied in 1920, are thin shadows of their original performances.[33] But the institute often kept or made copies of its recordings, and Hornbostel was a meticulous man. I do not think he meant to speak only of the thinning of wax. He also intended, I suspect, to say something meaningful about the process by which sound becomes a visible image of knowledge in the laboratory.

Written images were part of the psychological laboratory, thanks largely to the tools and instruments that researchers relied on. Like the notes of a song, beating hearts and twitching nerves were transient and rapid motions that defied capture, occurring in an instant and then lost. The ingenuity of nineteenth-century physiologists lay in the devices they designed to record and measure these fleeting moments. *Die Apparate*, the tools of the laboratory, included anything from a kitchen knife to a tuning fork, and sometimes even the animals of an experiment.[34] But the most intriguing *Apparate* were those specifically designed for experimental needs. There was the chronoscope, which could measure 1/1000 of a second with the flip of a lever. For acousticians, there was the Tonmesser, a calibrated pitch pipe that allowed its user to quickly determine the frequency of a given sound. And for psychologists, there was the ergograph, which isolated and measured the exertions of a muscle with a series of restraints and levers. The most valued all of these instruments, though, were the cylinders or *Registrirapparat* that laboratory technicians used to trace onto paper the vibrations coming from within a vivisected animal's body. The most widely used of these was the kymograph, a perfectly weighted rotating cylinder that, when attached to a stylus, could inscribe the infinitesimally small vibrations of a vivisected animal's circulatory system onto the turning surface of a blackened strip of paper.

As one visitor to Wundt's lab in Leipzig declared, the kymograph was "absolutely indispensable—more so than the chronoscope—to a laboratory in experimental psychology."[35] The instrument was a popular way to record blood pressure in early experiments with dogs (Figure 4.2), and quickly became used to document respiration, muscular motions, and movements of the heart, lungs, eyes, and other organs. The kymograph's inscribed graphic lines became the standard of physiological data, written information that had a one-to-one relationship with the body's motions. At one end, a pressurized tube received air, blood, or fluid from an animal's body, pulsing forward to raise and lower a disc attached to a thin rod. At the other end of the rod, a delicate stylus traced a curving wave on the blackened surface of the kymograph. In time, the instrument was adapted to record other types of vibrations in the laboratory, including sound. One of the first illustrations in Hermann von Helmholtz's famous book *On the Sensations of Tone* used an adapted kymograph to transcribe the vibrations of a tuning fork attached to a stylus, making visible the body of a sound wave.[36]

The kymograph provided other revelations about sound's inner workings. If you look at one of Thomas Edison's early cylinder phonographs, you will find that it is nothing more than a modified kymograph. Edison's machine worked backward: the rotating cylinder already had grooves in it, which caused the stylus to vibrate, shaking a delicate membrane that transmitted sound to a trumpet-shaped amplifier.

The prototype for Edison's phonograph was the phonautograph, first designed

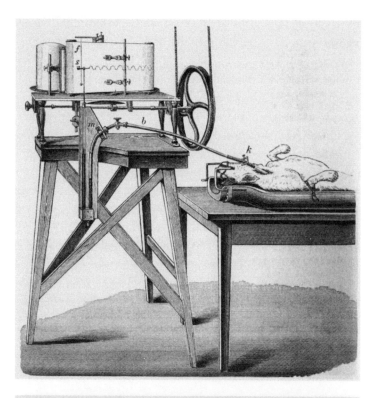

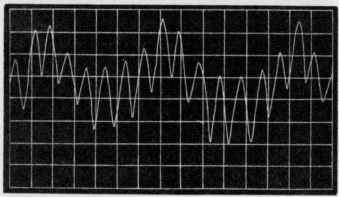

FIGURE 4.2 Kymograph used to measure blood pressure in a dog (*top*), and graphic result (*bottom*). As with other forms of vivisection, the dog dies during this experiment. Langendorff, *Physiologische Grahpik*, 206.

in the mid-1800s by the French inventor Édouard-Léon Scott. Scott's intent was not to record sound, but to make images of voices that would serve as a kind of stenography. His design worked more like a traditional kymograph: a voice would speak or sing into a trumpet-shaped amplifier, shaking a thin membrane that would vibrate in time to the sound waves, moving the edge of an armature that transmitted this up-and-down motion to a thin stylus, which, finally, would inscribe a graphic image of sound on a rotating cylinder. Scott's first attempt at a sound-writing machine was not much like a kymograph. Instead of a rotating cylinder, it slid a flat crystal plate directly under the trumpet-shaped amplifier. As the plate slid along, a boar's hair attached to the vibrating membrane at the end of the trumpet scratched an image onto the surface of the plate.[37] In 1859, Scott sent this design to the workshop of a well-known *Apparate* maker named Rudolph Koenig.[38] When the machine entered Koenig's workshop, it had the rectangular plate design; when it left, the plate had been replaced with the cylinder mechanism, clockwork, and armature of the kymograph.

Historians of sound recording have traced Edison's machine to Scott's phonautograph.[39] But Hornbostel appears to have noticed something self-evident to his contemporaries that historians have missed: the cylinder-and-needle combination of the early phonograph is an extension of the kymograph's technology. And with this technological ancestry, Hornbostel seems to have intimated a connection between sound's inscription in the grooves of the phonograph record and the ethics of knowledge that inscribed motion from the body of a living animal.

For Hornbostel, with one foot in the concert hall and the other in the laboratory, the relation between the kymograph, the phonautograph, and the phonograph would have been unmistakable. Figure 4.3 shows all three, first a kymograph, then a phonautograph, and finally a cylinder phonograph. Their contours offer a brief history of the transition of data from the body of animal to the production of sound. The technology of this transition remained remarkably stable: a rotating cylinder in which a critical inscription was stored, attached to a delicate vibrating stylus that made inscription possible.

The consequences of this similarity for the wax-cylinder animal appear serious, though less dire than the case of a laboratory animal. In the vivisection of a wax cylinder, the laboratory animal is replaced with a song. Notes, rhythms, and orchestration replace blood flow, respiration, and anatomy. But with this exchange comes a loss of data. The direction of motion in the phonograph is reversed from that of the kymograph: the grooves of the cylinder vibrate the stylus, which vibrates the membrane and trumpet, which amplifies the sound. In

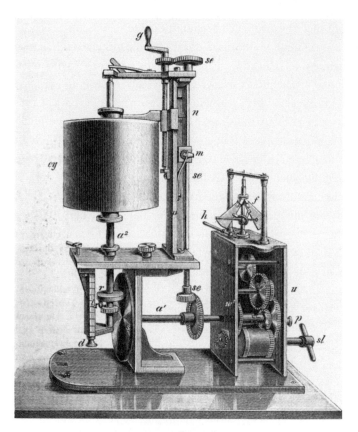

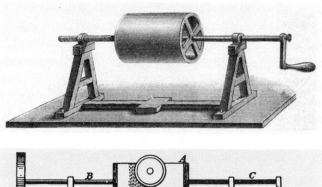

FIGURE 4.3 *Top*, kymograph, showing clockwork mechanism (Langendorff, *Physiologische Grahpik*, 19). *Middle*, Scott phonautograph (Langendorff, *Physiologische Grahpik*, 30). *Bottom*, diagram of a cylinder phonograph (Gillett, *The Phonograph, and How to Construct It*, 10).

this reversal is lost the kymograph's primary attribute, that of a two-dimensional visual image of motion in time.

Was it this transformation of living music into a written image that Hornbostel had in mind when he advocated the vivisection of a song?

To answer this question, I turn now to "Thai oi Kamen."

## UNPICTURABLE

Writing is a strange thing. The pictures made up of lines and curves that you are looking at right now are full of meaning. But while the image is visible, its meaning is not. "Unpicturable." Dr. Hornbostel called the meaning of "Kham hom" and "Thai oi Kamen" "unbildlich," *unpicturable*, when he approached their vivisection. It was not notes he wanted, though it would seem necessary to him to cut them away from the melody in transcription. It was meaning.

"Unpicturable: We need notes, which are certainly 'audible' to the eye; but there are also marks and schemas that are not. And if the analyzing intellect reveals wonderful laws and rules in this way . . . is it not conceivable that the creator found them too without any writing?"[40] These words follow the gruesome metaphor of vivisection in Hornbostel's writings. With them, he explains that the purpose of analyzing "Thai oi Kamen" was to discover these unpicturables. His goals echoed the distinction between anatomy and physiology, for the anatomist merely describes a body or a song, whereas the physiologist seeks to discover its internal functions by cutting open the body and seeking within.

In search of "Thai oi Kamen"'s unpicturables, Hornbostel began with a kind of gross anatomy, a version of the dissection with tonometer and metronome that he had recommended in 1905 for the songs of Hottentots. He first searched out the recording's life history: the song "Thai oi Kamen," or "Sad Departure," had been recorded in 1900 by Stumpf, during a Siamese theater troupe's visit to Berlin. Photographs of the troupe and secondary recordings fleshed out this history. Hornbostel then named the recording's instruments: the ranat, the kong yai, and others, many of which were still in Berlin. With these instruments and the handful of recordings made by Stumpf, the arduous and difficult process of finding out the song's physiology began: the first step, one might say, of the vivisection proper.

Sitting in the confines of the Phonogramm-Archiv, Hornbostel would have held up to his ear a Tonmesser or one of the carefully calibrated tuning forks ordered long ago from Paris for the Berlin Psychological Institute. Before him, the wax cylinder would rotate in the phonograph's scaffold, the delicate stylus lowered to its surface. With the help of these instruments, Hornbostel could confirm his ear's impressions and painstakingly identify the frequency of each note in the song. He would likely have written down his findings on scrap paper, and then rewritten them more carefully on one of the Phonogramm-Archiv's specially printed tables, as he slowly reconstructed "Kham hom"'s musical scale.[41] These dissected scales were among the Phonogramm-Archiv's most important products, and were created for as many songs as time allowed (figure 4.4).

With this information in hand, Hornbostel would have turned to a more extensive problem of writing: transcription. Hornbostel's published analysis of "Thai oi Kamen" included a meticulous set of transcriptions that used many of the annotations standardized at the Phonogramm-Archiv a decade beforehand. In 1909, Hornbostel and his coworker Otto Abraham had published guidelines that adapted music notation to the needs of scientific transcription, suggesting over forty ways to modify the traditional Western score in order to make it more precise.[42] But Hornbostel's transcription of "Thai oi Kamen" went far beyond a simple notation, for it was a structural analysis, comparing and assessing the use of phrases, rhythms, harmonies, and timbres in the piece. He mapped the relationship of tonics, dominants, and leading tones to repeating thematic material, drawing out the "now-intelligible whole" of his vivisection using the same language of symbols, charts, and graphs he deployed in his notes. The resulting images made music's otherwise invisible physiology visible through symbols, charts, and numbers.

Writing and its unpicturables were clearly important to Hornbostel and his notion of a scientific ethic. His process led toward the "laws and rules" of a culture that only intelligent reflection, the vivisector's moral compass, could reveal. It was exactly the kind of link between experimental psychology and cultural research that Stumpf imagined for the institute.

Other scholars seem to have independently adopted the ethic and method of intelligent reflection in their research as well (though without adding the frightening word "vivisection"). Hornbostel's student Jaap Kunst, in a study that made him famous, used a virtually identical method of extracting scales from recordings and instruments in order to compare the music of Indonesia to that of Central Africa. Like Hornbostel, he measured tones, recorded them in a table, and then sought in the resulting scales an invisible structure that connected the two cultural geographies.[43] Alexander Ellis, the British scientist who first translated Helmholtz's works into English, made a similar comparison between Scottish and

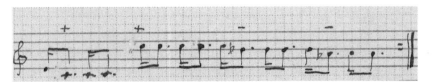

FIGURE 4.4 Hornbostel's preliminary study of "Thai oi Kamen" probably looked something like this study of a cylinder titled "Weii." Hornbostel took notes on scrap paper (*top left*), recorded his findings on the Phonogramm-Archiv's specially printed tables (*top right*), and adjusted his transcription for non-diatonic pitches (*bottom*). Charles Meyer, recordist; Borneo. Sarawak, "Weii," 1912–13. Max Wertheimer Papers, New York Public Library Box 11, Correspondence from E. M. von Hornbostel.

Persian melodies.[44] The results of these cultural comparisons were twofold, for in revealing a song's inner physiology, scholars also hoped to uncover the larger laws that connected culture to culture, and even species to species. These were, in effect, examples of the laboratory practice designed to confirm and expand upon the institutional typologies of chapter 2.

The songs of animals were also studied in this way. Ornithologist Aretas

Saunders used a graphic method to make visible the internal nature of birdsong "types," and discussed the merits of such typological graphing extensively with Wallace Craig in the 1930s.[45] Albert Brand, whose birdsong recordings formed the foundation of Cornell University's new ornithology program, used machines to transcribe his recordings in what he considered a more objective, scientific manner.[46] But the resulting images, Craig wrote privately to Brand, were weighed down by laborious details and didn't represent sound as it was heard by the human ear.[47] This complaint echoed Hornbostel's own desire for unpicturables, though it was written a decade later.

Another American naturalist, Sydney Ingraham, performed a melodic analysis in 1938 that was strikingly similar to those done by Kunst, Ellis, and Hornbostel. Figure 4.5 shows Ingraham's 1938 analysis of an American bird called a brown thrasher. First plotting the bird's cries on a graph, Ingraham proceeded to extract an arithmetic scale from the melody of the bird's cry. He then compared that cry to the slurred cries of primitive human music, situating the bird within a map of music history that extended from the thrasher to the well-tempered scale of Bach's day.[48] With Ingraham's graphic representation, we return to the lines of the kymograph, the inscription of sound, and the ethics of difference that permitted the laboratory animal.

Though such graphic inscriptions were not part of Hornbostel's "Thai oi Komen" analysis, they were highly valued images of musical meaning among Hornbostel's milieu. Frances Densmore, Benjamin Ives Gilman, and other ethnographers all used such graphic images. Their images, like the kymograph's inscriptions, were images of difference, plotting the songs of Sioux Indians, Hindus, Russian peasants, and other "others." Where the kymograph inscribed data directly from within its source, most studies of music had to painstakingly graph music by hand in order to recover the sounds of the phonograph. There were, however, attempts to return the process to its original integrity. Dayton Miller, an American physicist, developed a method of photographing light reflected in a mirror that vibrated in time with the phonograph's stylus.[49] And Carl Seashore, at the University of Iowa's psychology department, oversaw a number of experiments to transcribe sound, including attaching a phonograph to a kymograph, a process that reversed the lost inscription that the phonograph had incurred.[50] His instruments were used in a number of studies that described non-Western and nonhuman sounds including the songs of birds, African American folk songs, and Tlingit Indian language.[51]

Whether drawn by hand or inscribed mechanically, graphed notation was

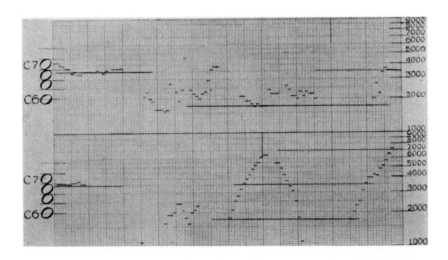

| | | SCALES | | | | |
|---|---|---|---|---|---|---|
| *Arithmetic* | | *Old Diatonic* | | | *Tempered* | |
| ratio | in 8ths | ratio | | | or logarithmic | |
| | 256 | | C | 256 | C | 256 |
| | | | | | C ♯ | 271.2 |
| 9/8 | 288 | 9/8 | D | 288 | D | 287.3 |
| | | | | | D ♯ | 304.4 |
| 10/0 | 320 | 5/4 | E | 320 | E | 322.5 |
| | | 4/3 | F | 341 1/3 | F | 341.7 |
| 11/8 | 352 | | | | | |

FIGURE 4.5 "This slur is the best proof of an arithmetic scale in bird song that I have so far found." Ingraham, "Instinctive Music."

a mark of the music scholar's knowledge within a scientific milieu. It was also, like the lines of the kymograph, a measure of difference. Such graphic images held a special fascination for Hornbostel. His notes are filled with them (Figure 4.6). Curved traces grow thick and hug ledger lines, push a melody onward with imperative arrows, or hover like birds on the wing in a blank page. They mark out the diverse and exotic sounds of an Indian raga, a Japanese koto, a song from Papua New Guinea. Such images were, Hornbostel wrote in 1909, not always precise. But they held within them the "psychologically meaningful subjective impressions" of sound.[52]

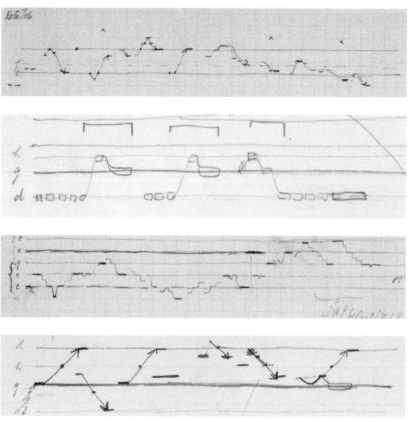

FIGURE 4.6  Musical graphs by Hornbostel. Max Wertheimer Papers, Manuscripts and Archives Division, New York Public Library. Astor, Lenox, and Tilden Foundations. New York Public Library Box 11, Series XI, c. 1912–1913.

## POSTSCRIPT

Let me take a moment to step back from these curving lines, and consider with you the question of this chapter as it stands. It begins with an individual, the man named Erich Moritz von Hornbostel, and the songs that were matters of his heart. He brought before us a proposition that appears to have been familiar to him, to his friends, and to his colleagues, though it seems strange today: to take those beloved songs, place them on the table of his laboratory, and kill them.

There seem to be two (at least) things at work here. One is an ethics of intelligent reflection, which psychological researchers inherited from Germany's physiological laboratories. This ethic was premised on the exchange of nonhuman life—life that is unlike our own—for human knowledge—knowledge of our own kind. Entwined with this ethic was the instrument through which it was implemented, the kymograph, and its offshoots, which included the phonograph. With these devices came images that represented hidden, invisible realities in visible and audible form.

Together, the ethics and instruments borrowed from German physiology gave studies of music a standard of data, of evidence. This data was grounded on notions of difference, the justification that allowed an exchange of animal/musical life for human knowledge. It was expressed in a visual language of unpicturables, the symbols, notations, and graphs that communicated not only information, but meaning. Data was thus *the written result of the exchange of difference*—life for knowledge, unpicturables hidden in coursing veins, pulsing lungs, and the quivering notes of a Thai orchestra.

Historians have long located the modern study of culture within transformations in German scholarship at the turn of the twentieth century that helped lay the foundations of the modern social sciences.[53] In Berlin, the stakes of this transformation took shape around an apparent tension between physiology's mechanistic tradition and the holistic, aesthetic approaches in Stumpf's institute.[54] But the story that I have drawn from Hornbostel's work is not primarily about the emergence of the social sciences or German holism. It is about the distant reach of a change in what counted as evidence, in standards of objectivity, that grew out of contemporary valuations of animal life. Evidence in Hornbostel's laboratory could be said to be about many things—mechanical devices, graphic imagery, tuning forks, or sound recordings. But the common point of origin for these techniques was the disposability of animal life, which had made it possible for European scientists of the nineteenth century to redefine quantitative standards of knowledge.

I argue that it mattered to Hornbostel, and still matters to us, that cultural knowledge traces its lineage to such a value judgment. A knowledge grounded in and by the lives of lesser beings brings with it a complex and extensive baggage. That baggage includes assumed oppositions between subjective and objective perspectives; between intelligent reflection and instinctive emotion; and between the value of others' lives and the value of knowledge. Those oppositions, in turn, are part of a far-reaching tradition of quantitative knowledge that was extended to the study of music. My purpose is not to say that notions such as "objectivity" are without meaning, but that their meanings are laden with long-forgotten assumptions and value judgments about animality. Those judgments continue to be important, for we continue to construct objective, measured, quantitative knowledge in relation to the animals, emotions, peoples, and sounds that constitute the sphere of subjectivity.

This chapter is a curious history of data—what is called an "epistemology," I suppose. Is the creation of data that comes from the disposability of animals moral? That is not epistemology, but ethics, which have a place in this story as well. There are many possible answers, of course. We could agree with Dr. Cullen that a dog's life is less valuable than the existence of practiced surgeons, and that our emotions will not guide us to moral decisions. We might, like Hornbostel, claim that musical knowledge is more important than musical integrity. Or perhaps we would provide the counter-answers: the dog is more important than surgeons; music is more important than knowledge; and subjectivity is undervalued. But these answers miss the mark. They accept the premise of the original debate, which is that our role is to weigh the value of difference against the value of knowledge.

Instead of accepting that premise, I would like to conclude this chapter by turning to a postscript of sorts and follow Dr. Hornbostel to his own experience of the logic of difference. The year after Hornbostel published his vivisections of "Kham hom" and "Thai oi Kamen," Stumpf retired and was replaced by one of his students, Wolfgang Köhler. Köhler returned to Berlin after research on primates in the Canary Islands that turned apes loose from the dissecting table to study their behavior, with surprising results that suggested high levels of intelligence. His was a directorship that began by questioning long-held assumptions about the differences separating human and animal minds. From this starting point, the Berlin Psychological Institute moved again from its spacious home on Dorotheenstraße into an even larger space, forty rooms in the old Imperial Palace. The Phonogramm-Archiv remained part of the institute, but it was also given more space, first in the Berlin Conservatory and later in its permanent home in the Ethnological Museum.

As these auspicious changes occurred, however, the Berlin Psychological Institute began to experience pressure from growing antisemitism in Germany. By 1920, when Hornbostel so vividly imagined placing a song on his dissecting table, debates about vivisection began to resurface in Germany in a new and disturbing form. In pamphlets and essays, the vivisector began to appear in the openly antisemitic form of a kosher butcher.[55] The growth of Germany's animal rights movement in the 1930s was accompanied by these representations of a

Jewish vivisector who gloried in animal cruelty, and was less than an animal himself. With this shift, notions of scientific morals grounded in human/animal differences were complicated in Germany by the insinuation that racial differences outweighed species boundaries. For Hornbostel and his Jewish colleagues in science, this insinuation added suggestive layers to the "intelligent reflection" that was valued at the Berlin Psychological Institute. Such reflection was indeed badly needed in Berlin.

Many of the institute's best-known members were Jewish, including Stumpf himself, as well as Hornbostel. Köhler, the new director, strongly opposed any attempt to dismiss his Jewish colleagues. His name is attached to the last anti-Nazi publication printed in Germany before World War II, "Gespräche in Deutschland"; but his influence was limited.[56] Only a few short months after the Nazis rose to power in 1933, Hornbostel was dismissed from the University of Berlin. He fled to Switzerland and then London, with plans to go on to New York and join the New School for Social Research. But amid these moves his health declined, and he died in 1936 before leaving for the New World.[57]

At the end of Hornbostel's life, suffering and scientific knowledge converged in the intrusive presence of racist ideology. Is this the final expression of difference toward which the kymograph pointed with its shifting stylus? Was the study of beauty in truth nothing but an expression of bias, racism, and cruelty? These were questions that psychologists, musicians, and biologists faced in the unexpected denouement of the 1930s. Their answers, which will be explored in the coming chapters, strove to escape this past. But the lines of the kymograph, and the inscription of sound, continued to chart the paths of knowledge.

# POSTMODERN HUMANITY, SUBJECTIVITY, AND PARADISE

———

In the final three chapters of this book, I trace the development of a disciplinary division that separated the study of human song from the study of animal voices. In that division, opposing standards of evidence emerged that located human and animal voices in separate spheres after 1945. In the study of animal voices, evidence was shaped by notions of laboratory objectivity, while in studies of human song, evidence was evaluated in relation to subjective cultural traditions. This growing divide between animal nature and human culture, however, was defined in many ways by concerns about the concept of race, which continued to trouble beliefs about genetic and cultural difference in the aftermath of World War II. Taken together, the last three chapters of my book tell a counter-history of humanism that revisits the problem of knowledge in a postwar, postmodern, posthuman world.

# FIVE

# Postmodern Humanity

A cold stone lion from Yugoslavia. Postal pigeons, carrying film during World War II. Bugs whose little black bodies are the letters of the alphabet, inspected by an illiterate child of India. The foot-and-mouth disease virus. Garuda, flying mount of Lord Vishnu and national emblem of Indonesia. A horse whose mouth is the grate of an American automobile. Palestinian sheep, bringing food to a desert wasteland.[1]

Here are the beginnings of a postmodern bestiary. Monuments, war, literacy, disease, national pride, the International Monetary Fund, and world hunger—such are the concerns of the world, filtered through the animals that appear in the pages of the United Nations Educational, Scientific and Cultural Organization, UNESCO, in the year 1950. One animal in this bestiary dominates all the others. It is a new kind of human, in whom body and mind are completely separated. One half of this human is pure biology. All of its physical attributes are encoded into chemicals that float through the innermost reaches of its body, passing on into the bodies of its children. The other half of the human is purely cultural, with its character, personality, and aesthetics stored in its mind, and passed on to the next generation through rich traditions and ceremonies.

Biological difference in the human animal, UNESCO assured its readers, had no relation to social, political, or moral life.[2] Though UNESCO's writings are a convenient reference point, descriptions of this human abound in the postwar era. Biologists asserted that genes and ethnicities were separate, while second-wave evolutionists in sociology and anthropology assiduously divided culture from biology.[3] Art historians argued that culture and genetics were separate elements of human experience.[4] For music's historians, the minds and cultures of humans were kept carefully apart from their biological bodies.[5] Wallace Craig's old classmate from Chicago, Oscar Riddle, complained about this new state of affairs where younger intellectuals refused to mix genetics with the assessment of mental characteristics.[6]

In this chapter, I explore the new human animal, a being created at the intersection of postmodern notions of human identity and postwar reactions to racial science. That topic is far broader than the histories I relate here, making this chapter a study of circumstance, of a set of intriguing coincidences that lay the foundation for the study of song in and after the 1950s. Here I offer three brief glimpses of the changing beliefs about human identity that circulated in scholarship after World War II in the fields of music, biology, and anthropology.

Postmodernism has been described by various authors as the spread of cultural relativism in the West in the second half of the twentieth century.[7] This chapter locates that shift through a parallel moment in the history of science, the evolutionary synthesis.[8] After a period of decline in the late 1930s and 1940s, evolutionary science underwent a revival in the 1950s as biologists turned to genetics and statistical models to develop stronger empirical approaches to the question of inheritance. Though the aims of the synthesis were framed by biological concerns, they were also shaped by a desire to distance evolutionism from prewar racial science. Classic descriptions of the synthesis tell of a neo-Darwinian unification of genetics and natural history, and often acknowledge the role of the synthesis in reviving the study of evolutionary science in biology.[9] The synthesis is not the subject of this chapter, but it serves as an important backdrop to my exploration of postmodern human identity.

Notions of the human in the postwar era were deeply informed not only by biology, but by changing attitudes toward race and its role in science. Horrifying accounts of the Holocaust circulated in the 1940s as Allied soldiers documented and publicized the discovery of German concentration camps. After the war, many biological and social scientists suggested that racial typologies had played a role in the rise of Europe's genocidal policies. Institutional conversations about race, however, existed in an uncomfortable counterbalance with mundane realities that restricted the rights and dignity of blacks, refugees, women, "third world" citizens, and others. Navigating the treacherous waters of history and politics, institutions like UNESCO turned to "scientific facts" as a reality beyond the reach of politics or culture. Those institutional narratives of postwar revision, change, and innovation located new ideas about humanity in relation to the Holocaust and the atomic bomb, advancing a history of the Holocaust, Nazi politics, and wartime destruction as the history of race. In the resulting divide between biology and culture, other histories of race and difference threatened to disappear as race came to be reimagined in the West as a political device exemplified by Nazi discourse.

Between those worlds of evolutionary science and racial politics, I trace the invention of the postmodern human animal through the various ways that music scholars, evolutionary biologists, and anthropologists thought about race in the years after World War II. I begin with the rise of "the music itself" in the 1950s, as music scholars attempted to excise evolutionary narratives from music's categories. From there, I turn to Charles Davenport's successor at Cold Spring Harbor, Albert Blakeslee, and the role of genetics in rebranding American evolutionary biology as a legitimate field of study in the wake of racial science. Finally I turn to the problem of race itself, and the way that biologists and anthropologists reimagined race after World War II, returning to the authors of UNESCO's human animal. This is a wide-ranging scope for a short chapter, and my goal is to be neither comprehensive nor complete, but to lay the foundations of the chapters to come.

Several years ago, historian of science Jenny Reardon introduced the connections between racial science and modern genetics by asking, "Who can speak for humanness? What is to count as a fundamental human science?"[10] As UNESCO's animal tells us, human identity was at stake in the reconfiguration of race after World War II. Though in this chapter I write of race, I am also thinking through race of the broader taxonomy of difference that was exemplified by species boundaries. I therefore begin this chapter with a paraphrase of Reardon's question: Who can speak for human identity in a society of racial violence? What is to count as postmodern identity? This is a chapter about the construction of the post-race human animal, and the price of memory that brought about its existence. Before World War II, music's science unified natural history with the comparative sciences of zoology and psychology. In the pages to come, I follow the rupture of those threads from previous chapters as these disciplines become increasingly segregated.

## THE MUSIC ITSELF

If you picked up a book of music history in 1955, you might read something like this: "The principal use of any book about music is to serve as an introduction to the music itself, which must always be the central object of study."[11] The author, Donald Grout, would tell you that the prelude to *Lohengrin*'s third act is often played separately as a concert piece, and that Richard Strauss converted to a Mozartean language in his later years.[12] But Grout would not talk about politics, or nationalism, or race. It was a stark refusal, and a contrast with books from the years before the war. In those old books, you could find a parade of singing birds, howling primitives, and cold Lutheran winters whose severity explained Bach's frigid fugues.[13] You could even be persuaded, in that mélange of ontology, that Gregorian chant was explained by a picture of Satan and his demons puking up Jewish bodies.[14]

But postwar historians wanted nothing to do with the birds and the Jewish bodies and the evolutionary winter. For them, music was at the center of its own history, a trajectory where "music, growing out of its own past, has shaped its own development."[15] Changes in music's tradition were not caused by physical bodies, or nations, or tribal impulses; they came out of music's own styles, without needing external reference. The new history was called "style criticism," a history of "the music itself" that eschewed racialized discourse for analytic detail.

Grout's textbooks were the exemplar of style criticism, an alternate universe in which history's actors were not animals, races, or nations, but cantatas, symphonies, and operas. A professor at Cornell and two-time president of the American Musicological Society, Grout promised his readers that he would focus on music and avoid "gossip" about politics, race, or gender.[16] American racial politics or the Holocaust had no place at all in such a book, and the timely question of Wagner's influence on Nazi ideology was delegated to a footnote. The main text said simply that "it" (without explaining what *it* was) was not within the scope of a book about the "art work itself."[17]

By 1960, Grout's approach was consolidated in a new textbook, *A History of Western Music*, which explained that the "history of music is primarily the history of musical style, and cannot be grasped except by first-hand knowledge of the music itself."[18] Now in its ninth edition and one of the most popular music textbooks in the world, *A History of Western Music* became the gold standard of music historiography. Reviewers praised it for its "objectivity" and its emphasis on the "purely musical side of music's development."[19] Not everyone agreed with Grout's new trend—one of the most vocal critics of music's evolutionary history, Warren Dwight Allen, objected to the divorce between music and its social contexts in which tunes "like Topsy, 'just grew.'"[20] But for the most part, "the music itself" was praised for its potential to unify different pieces of music from different times and places through a coherent story about changing styles and genres.[21] Successful scholars adopted the approach, and music's history became a history of style.[22]

Where did this wholesale suppression of music's scientific past come from? How was it possible? Music was not what it once had been. The younger generation spent less time with sheet music and musical instruments than their elders did, relying instead on the radio and phonograph records. They listened less to classical music than Davenport's generation had, preferring popular tunes and jazz mined from an increasingly global market.[23] Governments tried to promote Cold War agendas through music, often with mixed results, as when Gershwin's *Porgy and Bess*—a jazz opera set in an impoverished Southern community—was sent abroad by the US State Department to combat the reputation of American racism in the 1950s.[24] In places like Cold Spring Harbor, where Davenport had once stashed a huge chart of the Bach family tree to show how musical talent was inherited, music had lost some of its cultural capital.[25] The new director at the labs was uninterested in music's science, limiting himself to classical records and the popular tunes he whistled for his dog, Prince.[26]

In this world of gently fading masterpieces, a line began to be drawn between the study of music and the study of evolutionary science. Students of music's history were reminded in the 1940s that there was a "fundamental difference between the idea of evolution in history and in natural science."[27] One music historian wrote that evolution was "hazardous," asking his colleagues not to confuse music's development with the biological evolution of a species.[28]

Some blamed German empiricism for the collapse of their cultural heritage. "Germany is now dead," wrote one music critic stunned by the revelation of the Holocaust, believing that it was the fault of the "dry empiricists" who had made music and the arts a "dissecting table."[29] The collection and classification of songs lost much of its appeal after the war for the younger generation; by the 1960s, newer scholars were repelled by the idea of "making taxonomies, . . . cutting and splitting, drawing boundaries, differentiating between one thing and another."[30]

The song collectors took their cue from this changing world. By 1950, the language of collection and comparison began to be replaced by terms such as "ethnographie musicale," "ethnomusicology," and "Musikethnologie," aligning studies of non-Western music more strongly with cultural anthropology than with physical anthropology or natural history.[31] Song collectors like Laura Boulton and laboratory investigators like Hornbostel became "ethnomusicologists." A handful of postwar ethnomusicologists became second-wave social evolutionists, dividing race from culture by making the distinction between "socially transmitted" musical evolution and the "biologically inherited" patterns of genetics.[32] Others rejected evolutionary metaphors outright, associating them with what Hungarian folklorist Constantin Brăiloiu described sarcastically as faith in "the particular genius of 'Whites' . . . the measure of all things musical."[33]

Writing in his New York apartment in the 1950s, Curt Sachs was one of those music scholars rethinking evolutionary hierarchies. Like Hornbostel, Sachs had been ousted from his teaching post in Berlin after the rise of the Nazis, and fled to Paris where, as described in chapter 2, he had wryly imagined giraffe-pianos stretching their necks behind the glass cases at the Musée d'Ethnographie by the Seine. One imagines that those giraffes made him sad in the war's aftermath. After dedicating his career to understanding music's evolution, Sachs believed he had earned the hard-won insight that rigid typologies of race, the cornerstone of his life's work, were to blame for the war's crimes. Instead of a parade of giraffe-pianos whose oddly shaped necks revealed the progression of national styles, he now saw a treacherous web of cultural complexity.[34]

Like so many other music scholars, Sachs wanted to leave behind racial typologies and linear evolutionary theories for a more limited scope. His postwar prose reveals an uncharacteristic insecurity, describing musicologists as ill equipped to offer judgment on larger issues.[35] Many scholars of the postwar era, both in music and in science, seem to have avoided racial questions in part from a genuine sense of their own inability to provide answers. Sachs didn't expect his own work, or "the music itself," to resolve the problem of what race was, and how or if it related to music.

Instead, he expected genetics to solve it. "The criminally exploited concept of 'race,'" he wrote, was the result of confusing "biologically transmitted genes" with "environmental culture," spheres of biology and cultural study that, he implied, should remain forever separate.[36] This momentary turn to genetics reveals a whole world of heredity that, when Sachs first published his taxonomy of musical instruments with Hornbostel nearly half a century before, simply did not exist.

## DATURA STRAMONIUM

Sachs's faith in genetics sends me to Cold Spring Harbor in the late 1940s, when Albert Blakeslee, a botanist affiliated with the laboratory, used the same division between genes and culture to defend his favorite plant. His letter was addressed to the poet Edna St. Vincent Millay, who had asked the unflattering question,

> Here the rank-smelling
> Thorn-apple,—and who
> Would plant this by his dwelling?

The naming of this plant, known to Blakeslee as *Datura stramonium*, was redolent of the confusion that prompted botanists to adopt the type specimen at the turn of the century. In England it was called "thorn apple" and "raving night shade"; in the United States, "Jimson weed" or "stinkweed." The French called it *pomme épineuse* and *herbe aux sorciers*, the Germans *Dornapfel*. To the Russians, it was a word that meant "stinking stupefier," and to the Japanese, *yoshu*, or "foreign."[37] Its pale white flowers are lovely, but the plant is poison and its seeds are cased in thorns. Sanford's *New England Herbs* stated that it could be found in "almost every dump heap and piece of waste ground near large cities," while Kearney and Peebles called it "coarse" and "ill-scented," noting ominously that it had poisoned children, cattle, and sheep, and adding with dismay that Indians allowed their children to ingest small amounts as a narcotic—"a dangerous practice."[38]

To Blakeslee, the plant was the icon of modern genetics. "*I* would plant this by my dwelling, and have done so for the last thirty years rather extensively," he wrote firmly in his missive to Millay. "It turns out that this plant (*Datura stramonium*) is perhaps the very best plant with which to discover principles of heredity."[39] Blakeslee's letter to Millay came at a moment when the principles of heredity were under careful scrutiny. Three years before UNESCO promoted the new human animal, questions about breeding and racial difference had become treacherous ground that threatened the foundations of evolutionism. Holding up the chromosome in defense of the stinking, foreign, dangerous sorcerer's herb, Blakeslee saw *Datura* as a beautiful plant because it taught the world with greater clarity what genes could—and, by implication, could not—deliver.

Blakeslee, who served as the director of Cold Spring Harbor between 1934 and 1941, defended the lab's reputation when he defended *Datura*. Cold Spring Harbor had fallen on hard times in the days since Laura Boulton had worked there.[40] When Davenport retired in 1934, the three-part complex of summer school, genetics laboratory, and Eugenics Records Office was ailing. The lab's eugenics research, after a peak of popularity in the 1920s, was dragging down the reputation of the entire complex.[41] While wartime funds poured into nearby engineering projects at Bell Laboratories and the headquarters for the Manhattan Project in New York City, Cold Spring Harbor's lab was so hard up during the war that historian Nathaniel Comfort describes the director after 1941, Milislav Demerec, wandering the grounds to find unused light bulbs, unscrewing them and saving them for later.[42]

Evolutionary science in general was suffering from its associations with eugenics and racial science. Historian Vasiliki Smocovitis describes how introductory biology students at Harvard were told in the 1930s, "Evolution is a good topic for the Sunday supplements of newspapers, but it isn't science."[43] By the 1940s, that Sunday-paper aura called forth bitter memories of eugenics and racial science that had become irretrievably associated with Nazi politics. Rumor had it that academic biologists in Germany had joined the Nazi party at a higher rate than members of any other profession,[44] and biologists from other nations were beginning to associate prewar evolutionism with Hitler and the Nazis.[45] These were, Blakeslee confessed in 1940, "dark days" for science as he knew it.[46]

Cold Spring Harbor responded to these changes by slowly rebranding itself as a center for modern genetics. This was an area that seemed to be gaining traction in the first half of the century. There was the groundbreaking study of *Drosophila* in 1910 at Columbia directed by Thomas Hunt Morgan, and new genetic inroads in human medicine followed with Ronald Fisher's work in the 1920s.[47] At Cold Spring Harbor, Davenport's successors were invariably geneticists: Blakeslee took over from Davenport in 1934, and was succeeded in 1941 by the geneticist Milislav Demerec, a lineage that continued with the appointment in 1968 of James Watson, one of the discoverers of the DNA double helix. During the 1930s, Blakeslee gradually shut down the Eugenics Records Office,[48] and with Demerec's help, Cold Spring Harbor became a center for international conferences and meetings at the forefront of modern genetics.[49] Blakeslee hoped to separate the lab's science from ethical debates, telling a class of college students in 1940 that science was blameless in war, for it was "the wayward mind of man" that caused human atrocities.[50]

While music scholars like Sachs blamed empiricism for racial science, scientists embraced empirical research as its antidote. When Blakeslee had started his studies of *Datura stramonium* at Cold Spring Harbor in 1915, a gene was a thing, a singular unit of biological inheritance that, like a mail carrier, delivered a trait to your body all at once. By 1955, the mail carrier had morphed into an entire community of molecular chemistry embodied in DNA's double helix. The gene became nothing more than a laboratory experiment, a little tampering with chemicals in a petri dish that had a consistent effect on the unfortunate chromosome's offspring. Instead of being tidy units, traits were now the result of scattered and unpredictable contributions from this biochemical community. Brown-eyed dominance remained delightfully predictable in textbooks, but many other traits disappeared into an indecipherable chemical mire. From the viewpoint of genetics, racial traits and types had been an abusive misunderstanding of this biochemical complexity.

Blakeslee's own work focused primarily on chromosomes, the gene's carriers. *Datura*, which had plentiful seeds and a rapid life cycle, was his passion, rumored to have a place in his heart between his first and second wives.[51] It seemed ideal for chromosome research, and Blakeslee and his colleagues rapidly produced it in vast numbers. Cold Spring Harbor's greenhouses, originally used by Blakeslee's predecessor George Shull to hybridize corn and to grow winter lettuce for the lab's animals, had become an impressive complex situated near the Animal House and the property's dormitory. In the summer, the nearby grounds were speckled with tents that popped up to accommodate women who, like Laura Boulton, did clerical work on a bare budget.[52]

Breeding *Datura* in these greenhouses and then examining histological sections under a microscope, Blakeslee and his team discovered some surprising things about heredity. They identified at least five hundred forty-one mutations of *Datura*, mapping many of those changes to specific chromosomes and providing the bare outline of something like a *Datura* genome in the process.[53] Blakeslee's team members at Cold Spring Harbor found they could make the chromosomes mutate faster if they x-rayed the plant or brushed it with colchicine, creating what must have felt like mini-evolution in mere months. The results were impressive, and inspired later work with genetically modified flowers and crops.[54] The team made plants with twice as many chromosomes as they needed; chromosomes with extra or missing parts; and even plants with mixtures of normal and modified chromosomes that could be followed like a trail of breadcrumbs, to figure out which kinds of cells—and which hereditary instructions—ended up in leaves, stalks, or roots. In the midst of all of these mutant plants, inheritance began to take on an incredibly specialized, detailed, and concrete character.

Genes and chromosomes, which provided a new image for Cold Spring Harbor, were also the face of the lab's new annual symposium series. The symposia, run by Blakeslee's successor Milislav Demerec, offered another noticeable contrast to Davenport's Eugenics Records Office. Attracting leading figures in biology, the lectures gave the laboratory a name for empirical research at the forefront of biochemical studies of inheritance. The series had been canceled during World War II due to budget shortfalls and the impossibility of including scientists from other countries. But it was resurrected in 1946, and gave the labs a venue to field new theories in biochemical evolution. The topics were concrete, and the scope was clearly defined: Heredity and Variation in Microorganisms (1946); Nucleic Acids and Nucleoproteins (1947); Biological Applications of Tracer Elements (1948); and Amino Acids and Proteins (1949).

But the session in 1950 was special. Instead of focusing on genetic biochemistry, Demerec allowed a geneticist and an anthropologist, Theodosius Dobzhanzsky and Sherwood Washburn, to plan the year's meeting around an ambitious topic: Origin and Evolution of Man. The purpose of the conference was to establish collaboration between biology and anthropology, and, less explicitly, to reclaim the central role of evolution in the sciences.[55] Understanding humanity's "place in the universe," wrote Demerec, was the job of all branches of science, but of evolutionary anthropology and biology most of all.[56]

The symposium brought together a remarkable range of specialists in genetics, natural history, primatology, and physical anthropology. One of the meeting's tag words was "population," sprinkled liberally throughout Washburn's presentation on primate evolution.[57] Populations were a new way to think about human behaviors and natural selection, replacing the singularity of the type specimen with vast numbers that seemed to promise quantitative statistical analysis. By the end of the nine-day meeting, a picture began to emerge of a new kind of evolutionism, one in which naturalists, anthropologists, and geneticists cooperated to forge a unified evolutionary science that encompassed everything from molecular biology to human anthropology. Genetics, numbers, and data—exactly the kind of work that Blakeslee was doing with *Datura*—were going to save evolutionism.

## RACE, BIOLOGY, AND ANTHROPOLOGY

Origin and Evolution of Man would come to be perceived as one of the turning points in evolutionary science, part of a moment when the disillusionment of the 1940s gave way to a revival of Darwinian evolutionism shared by biologists and anthropologists. Many of the symposium's attendees were central figures in the growing unification movement called by its proponents the "evolutionary synthesis," and the meeting allowed them to promote and develop their ideas.[58] Darwin's legacy was the platform for the synthesists, who adopted a marriage of natural history and genetics under the same rhetoric of "populations" that Washburn used in his paper at Cold Spring Harbor.[59] Populations became the seat of species identity that had been reserved at the beginning of the century for a single type specimen. And the nature of that identity was quantified not through animals' appearance, but through their genetic makeup, placing the specialized details of work like Blakeslee's at the center of postwar practices of biological identification. As historian Vassiliki Smocovitis has shown, the synthesis was the revival that evolutionary science needed, bringing evolution back from its associations with speculation and racial science.[60]

Though the central aim of the synthesis was the unification of genetics with other branches of natural history, this was also a moment for biologists and anthropologists to respond professionally to evolutionism's heritage of racial science. In biology, advocates of the synthesis explicitly associated prewar notions of typology with racism, contrasting "types" based on singular type specimens with newer population genetics.[61] Two leaders of the synthesis (and speakers at Origin and Evolution of Man), the naturalist Ernst Mayr and the geneticist Theodosius Dobzhansky, were outspoken in framing typology as an expression of racial science. Mayr called typology one of the "crimes of the racists,"[62] while Dobzhansky wrote, "A type once created has an insidious way of dominating its maker. It becomes 'the race,' a sort of noumenon of which the existing individuals are only imperfect representatives . . . people with O blood group, or long-headed people, or criminals."[63]

The old practice of using type specimens and developing typologies, once the key to developing evolutionary narratives, was now "anti-evolutionary" for Dobzhanksy.[64] Shifting populations replaced types, and large-scale samples replaced the type specimen. Implied in this rejection of types was, among other things, a repudiation of Agassiz's school at the Harvard Museum of Comparative Zoology, including Cold Spring Harbor's founder, Davenport.

Anthropologists, too, revisited their relationship with racial evolutionism in the postwar era.[65] As in biology, prewar typologies and progressive narratives were dismissed, in the words of Margaret Mead, as "myths of Nordic superiority" that belonged with the "clipped hedges and rounded corners for nursery tables" left over from the nineteenth century.[66] Like musicologist Curt Sachs, anthropologists replaced these myths with a strict division between biology and culture. Second-wave evolutionists adopted a hard dividing line separating genetically transmitted inheritance in biology from the anthropological study of socially transmitted traditions.[67] As sociologist Morris Ginsberg explained, "we must, I think, accept the conclusion . . . that progressive modification of civilization which passes by the name of 'evolution of society,' is, in fact, a process of an essentially different character [than biological evolution]."[68]

Only weeks after the Origin and Evolution of Man symposium was held at Cold Spring Harbor, Ginsberg worked with Dobzhansky to draft a statement on race for UNESCO in Paris.[69] The UNESCO Statement on Race, issued a month after Origins and Evolution of Man, outlined a rule book for definitions of race in the postwar world. Its authors included symposium attendees Dobzhansky and Ashley Montagu, and drew on the premises of the synthesis. The result of their work was one of the most comprehensive descriptions of the human animal as it was imagined in the aftermath of World War II. In the UNESCO statement, human identity was a composite of biology and culture, relegated to separate spheres by the conundrum of race.

Race, in the UNESCO statement, was posed as a virtually impossible double reality: biologically true, but ethically unspeakable. Borrowing the language of the synthesis, the statement asserted race as a factually true category when properly defined "from the biological standpoint" in terms of genetics and populations.[70] But the document's authors also included a long litany of categories that were *not* race: Americans, Englishmen, Frenchmen, Catholics, Protestants, Moslems, Jews; people from Iceland, England, and India; people who are culturally Turkish or Chinese; and many more. Considering the many "serious errors" that had been committed under the rubric of race, the document's authors suggested that "it would be better when speaking of human races to drop the word 'race' altogether and speak of ethnic groups."[71]

In this world, race had become real but unspeakable. Like "the music itself," modern genetics rehabilitated human knowledge from its racist past. But the cost of that rehabilitation was an uncomfortable doublethink, a simultaneous acceptance of two contradictory beliefs about human nature.[72] Postwar intellectuals accepted genetic measures of race as real. They also accepted unequal measures of human cultural groups as unjust. But in isolating genetics from culture, it had become impossible to resolve the two worlds in which race was biologically real, and socially unjust.

The fabric of racial doublethink was ingrained in the spaces of Cold Spring Harbor and UNESCO's Paris headquarters, where compartmentalized approaches to race were part of a well-rehearsed social norm. In documents from Origin and Evolution of Man and UNESCO, racism was exemplified by the antisemitic descriptions of Jewish identity that preceded the Holocaust. Yet American race relations in the 1950s reflected anti-black social policies and legislation that cultivated habits of perceiving whiteness as neutral. Origin and Evolution of Man included no black participants among its hundred and twenty-eight attendees from wide-ranging countries, and only a handful of women, most of whom attended in a spousal role. Cold Spring Harbor was just a few miles away from Levittown, one of the nation's largest segregated suburban communities, and the lab's district reported only five nonwhite residents out of more than fourteen hundred in the 1940 census. In Paris, the UNESCO statement on race was likewise drafted by a small international collection of voices that represented neither the voices of black or Islamic authors, nor women. Yet the same decade was marked in French daily life by the collapse of French North Africa and a violent civil war in Algeria. My point is not simply that the architects of race in Cold Spring Harbor and Paris failed to engage with real-world racial politics, but that their world was structured to encourage that failure. The scope and context of biology's political context in the 1950s is far too complex to address in this chapter. But it is important to acknowledge that scientists' postwar discourse of race, which engaged directly with antisemitism and structured historic ruptures and changes around the Holocaust and the atomic bomb, occurred within contexts in which other racial histories were both urgently present and unrecognized.

In the sphere of biology, this doublethink was epitomized by work like Blakeslee's with plants. In agriculture, race was part of a tradition designed to define and measure ways to maximize food production. Like the Palestinian sheep in my opening bestiary, plants such as *Datura* were part of a broader conversation about how to feed a global population. At Cold Spring Harbor, this racial discourse was particularly salient in studies of corn. Blakeslee's predecessor at Cold Spring Harbor, George Shull, had crossbred hybrids of corn that were enormously successful. Paired with Blakeslee's studies of the effects of colchicine, corn research at the lab helped to lay the foundation for the later development of genetically modified breeds of corn. Before the war, breeds of corn were accepted as fundamentally similar to the notion of human difference.[73] After the war, it was better, as the UNESCO statement explained, not to speak of race at all. But geneticists like Blakeslee continued to be part of an active economy in which races of corn and other plants were imagined as biologically real and economically valuable facts. The result was a kind of double life, in which non-human biology continued to be premised upon categorical differences that were clearly comparable to prewar notions of race, while race in human biology was approached through rhetorical avoidance.

For scientists like Blakeslee, the result of this doublethink seems to be not an intense self-examination about the role of cultural hierarchies in scientific discourse, but an avoidance of comparisons like the corn/human analogy. Some scientists argued that the responsibility for antisemitism and nationalism lay at the feet of "laymen" who lacked a real understanding of evolutionary science.[74] Most, like Blakeslee, tried to separate science entirely from the problem of race.[75]

Demerec, whose symposia rebranded the labs as a place of modern genetics, tried to reopen the Eugenics Records Office in 1942.[76] At the 1950 symposium, four of the nine sessions were devoted to the subject of human racial difference, making almost half of the meeting—best remembered as a moment in the evolutionary synthesis—a discussion of the practice of collecting, classifying, and analyzing human difference.

Who could escape this world of doublethink? One rather curious solution was offered by an anthropologist named Wilton Krogman, who chaired a session at Origins and Evolution of Man but did not give a lecture. Krogman, a prominent professor of physical anthropology, was considered an expert on the scientific discourse of racial identity. But he was dissatisfied with explanations of race that were limited to primatology, paleontology, or genetics. Confronted by the apparent dichotomy between cultural and scientific explanation, Krogman came to believe that the rhetoric of race was an ethical dilemma, in which Nazi claims were being revisited in American race relations. Writing in 1948, Krogman attempted to redefine the problem of race as an ethical one with a turn toward biblical injunction:

1. Thou shalt not bow down before the false god of "racial superiority";
2. Thou shalt not vaunt thyself that only thy race is "pure";
3. Thou shalt not preach that races or peoples are at different levels of physical development;
4. Thou shalt not claim that racial differences are of fundamental import;
5. Thou shalt not establish racial groups as fixed or unchangeable;
6. Thou shalt not, to the detriment of thy neighbor, assert that cultural achievements are based on racial characteristics;
7. Thou shalt not hold that "racial personality traits" are innate and inherited;
8. Thou shalt not demean thy brother because his appearance differs from thine;
9. Thou shalt not, because a man is of a different religion, insist that he belongs to a different race;
10. Thou shalt faithfully and sincerely observe the foregoing admonitions and then, indeed, wilt thou love thy neighbor as thyself.[77]

The "son of Man," Krogman explained, could "know no arbitrary biosocial refinements that make for inequality" such as those evidenced in American race relations.[78] Krogman's human was not governed by the discourse of genetics, or immigration politics, or war, but by a binding concept of human dignity. To accept the doublethink of race was a "willful sensory denial" on the part of Americans who had chosen to blind themselves to race prejudice. "We did not wish to see or hear," Krogman wrote, this time paraphrasing the New Testament.[79]

## THE POSTMODERN HUMAN

As I come to a halfway point in this book, I return with Krogman's moral animal to the bestiary of UNESCO and the question "Who can speak for identity?" This chapter began with a refusal, the refusal of music scholars to confront more than "the music itself" in the aftermath of a difficult period of racial violence. A similar refusal accompanied the turn toward genetics at Cold Spring Harbor, where the past that had once connected music to racial evolutionism was gradually erased in the lab's attempt to redefine itself. These refusals to engage with race were surrounded by a number of complex motives: a desire, like Blakeslee's, to disassociate the field, and by extension, the friends and colleagues in that field, from the crimes of racism that were, by the 1950s, being literally prosecuted in the courts; or an uncertainty, like Sachs's, that the tools acquired before the war left one equipped to do justice to the challenges of the present.[80] It was in the midst of these motives that the individuals who reimagined human identity in the postwar era became preoccupied with the separation of biology and culture. I have suggested in the previous pages that the mechanism of that new model of human identity was a contradictory approach to race, in which difference was simultaneously accepted and unvoiced. I conclude my chapter by suggesting that the separate spheres of culture and science, which defined the special status of the human, were, in many ways, also the postmodern condition of race as it was understood by the biologists and social scientists to whom Krogman addressed his concerns. [80]

In the aftermath of World War II, the old science of identity based on comparative zoology and psychology seemed to come to an abrupt end, but persisted despite the synthesists' efforts. Music scholars remembered the past as a period of "comparative musicology," but forgot to edit old passages in their books in which German music was still affected by the enduring winter, Italians still wrote joyful melodies because of the local sunshine, and Jewish composers still produced trite theatrical songs for profit.[81] On Long Island, the librarians at Cold Spring Harbor who succeeded Laura Boulton stored the relics of the Eugenics Record Office in folders and boxes, eventually providing digital resources that preserve and historicize this difficult chapter in the lab's history.[82] Yet the lab also cultivates a complex counter-history of genetic innovation, and has worked to rehabilitate Charles Davenport as a founding figure of genetics and cancer research, rather than eugenics.[83] The result is a set of conflicting narratives in both music and science about race and its relation to social and biological classification. The conflicts that cultivate these dualisms might be summed up in UNESCO's human animal: uniquely unlike other animals in its culture and status; divorced from breed categories in name; experienced through racial typologies in practice. The omissions and conflicts that defined this creature, I suggest, are one way of understanding the concept of race as it transitioned from the first half of the twentieth century to the second half in a history whose periodization is itself defined by the monuments and landmarks of a white, European history of war and genocide.

Who can speak for identity? UNESCO's human animal, a postwar being uniquely divided between genetics and culture, helped scientists and scholars move beyond racial science. It shaped a much-needed path forward for both biologists and scholars of culture in a world that was fragmented by racial violence. Yet in making race unspeakable, the postwar, postmodern human also erased the traces of continuity that connected the practices of comparative science to newer methods and traditions. By dividing culture from nature, and by dividing history around the Holocaust and the end of World War II, Western intellectuals also divided themselves from the past. Though this chapter only hints at the causes and effects of that historical amnesia, it suggests an outline of the conditions under which music began to leave the sphere of scientific inquiry.

In the following chapters, I chase the concepts of identity and knowledge formulated in the first half of the twentieth century into the aftermath of this midcentury turn toward genetics. In postwar studies of song, the unique status of the human animal becomes a central feature of the ways that music is imagined and analyzed. There the postmodern conditions of race, which defined an unspeakable category of human identity in the 1950s, also become the conditions under which musical capability was defined. In the pages to come, I follow that restriction of musicality to the human animal into the postwar analysis of musical voices, and their use as sonic markers of biological identity. From the double lineage of comparative zoology and comparative psychology that lay the foundation of the first half of this book, I turn to the reforging of those traditions into the unrecognizably disparate worlds of modern ornithology and postwar music scholarship.

# SIX

# Listening for Objectivity

Being human, William Homan Thorpe and Charles Seeger were naturally suspicious of their own observations.[1] It was in 1950 that the two men decided independently to buy machines to replace their own ears in the laboratory. These machines would be proxy listeners, using electronic filters to transform recorded sound into a series of electric pulses that would, in turn, move a stylus, whose quivering needle inscribed an image on a nearby rotating spool of specially prepared paper.

To a casual observer, the machines looked like kymographs.

You might wonder why two trained human listeners would want to exchange their ears for machines in the first place. Thorpe and Seeger worked in different countries and in separate fields. They had never met and probably had never even heard of each other. Yet in the years after World War II, they shared a set of common concerns about music's role in the familiar path from professoriate to laboratory directorship. Thorpe, a deeply religious man and avid naturalist, was cobbling together a laboratory for avian song research on the outskirts of Cambridge, England, building dozens of birdcages from repurposed wire barricades left over from the war.[2] Seeger, a conductor and intellectual known for his brash confidence and administrative skills, was in the midst of a transition from government work to a post at the University of California, Los Angeles, where he would help develop a new institute for ethnomusicological research.[3] Without knowing each other's work, the two men independently came to the same conclusion: that studies of sound required them to doubt their own ears.

Hearing music in the 1950s was a disturbingly vulnerable experience. The decade was bracketed at one end by Theodor Adorno's claim that music manipulated the masses for political and financial gain, and at the other end by Pierre Bourdieu's discovery that musical preference had more to do with one's social class than with any intrinsic musical ability or taste.[4] Psychologists had argued for the past several decades that laboratory experiments showed listeners responding to factors that had nothing to do with acoustics.[5] In the aftermath of World War II, governments hoped to manipulate that vulnerability. The French inundated the radio waves of occupied Germany with French masterworks, while the Americans sponsored concerts they imagined would prevent the spread of communism.[6] Music scholars and scientists, meanwhile, were in the midst of the separation of musical knowledge from racial science described in chapter 5, with all its attendant anxieties about the misuse of cultural research in scientific settings.

In the laboratory, researchers needed some way to extract themselves from this vulnerable position. How could you speak for other animals, or even other people, when you couldn't be sure your own ears were reliable? Thorpe and Seeger found a partial solution to this problem in the sound spectrograph, a machine developed during the war by Bell Laboratories. Bell Labs had a long history of innovating ways to visualize sound, part of the company's philanthropic commitment to deaf education. During the war, the labs had turned to technologies like radar and sonar as ways to visualize the enemy.[7] The spectrograph, developed under engineer Ralph Potter, capitalized on the old fantasy of a voiceprint, an image engineers had once imagined would identify friend or foe "ten, 20, or even 40 years later."[8] One-half of the machine was a repurposed kymograph, replicating the form of the nineteenth-century laboratory apparatus in a rotating cylinder and stylus. The novelty of the new device was its other half, which had the capacity to strategically transform sound into a vibration that, once sent to the stylus, was easy to read. Using a series of electronic filters to screen out select bandwidths of sound from the source recording, the machine acted like a kind of prism for sound. The result was a compartmentalized vision of the original source recording, divided into intelligible information.

By transcribing directly from a source recording, the spectrograph replaced human listeners with a mechanical mediator that had a single, unchanging way of describing what it heard. The description itself, moreover, was a reassuringly familiar image borrowed from the kymograph's mechanics. Yet this process posed its own problems. Separating researchers from musical experience also separated them from music. Researchers found themselves in a double bind: to study music objectively, it was necessary to remove personal experience from the process of listening. But to remove personal experience was to remove music from sound. It was a partial reversal of the ethics of knowledge that had guided the prewar music laboratory, replacing the power of human subjectivity with a mechanical intermediary. In the past, the graphic image had been an unpicturable, a "psychologically meaningful subjective impression" constructed from the researcher's ability to generalize about music.[9] Now, the same image was an objective impression, created by a nonhuman machine that mediated sound.

In this chapter, I trace the introduction of that machine into the postwar music laboratory. Thorpe's and Seeger's work can be read as a counterpart to chapter 3, which described the work of two other song collectors, Wallace Craig and Laura Boulton. Like those older song collectors, one protagonist of this chapter is ornithological, while the other is anthropological. Many of the same aspirations and assumptions about field and laboratory informed their work. But unlike Craig and Boulton, Thorpe and Seeger occupied utterly separate spheres. My chapter follows these two postwar song experts into the divided spaces of zoology and musicology, where shifting beliefs about the human animal presented researchers with complex questions about machines, objective ears, and the making of musical evidence.

At the center of these conundrums was the notion of human subjectivity. Subjectivity is the main topic of this chapter, which turns to the mechanical transcription devices of the postwar era and the difficult choices that this new technology presented. Historians of science Lorraine Daston and Peter Galison have argued that subjectivity and objectivity are paired twins whose meanings fluctuate in relation to time and to one another.[10] In the aftermath of World War II, specialists in song were faced with a conflict that seemed to come from within, putting researchers' capacity for disinterested judgment at odds with their sensation of musical sound. It was a uniquely human problem, for researchers shared a faith not only in their hope of objectivity, but in their common capacity to hear music in a deeply personal and interior manner. Out of this uniqueness emerged a secondary question: How could it be possible for postmodern comparisons of human music to abide by the same standard of knowledge as that which guided studies of animal song?

## THE SPECTROGRAPH AND THE MELOGRAPH

The study of birdsong has its own family lore.[11] It begins in 1950 with a machine that transcribed sound into pictures at the Admiralty Research Laboratory, in the London suburb of Teddington. Historian of science Gregory Radick tells how for two years, the forty-eight year old zoologist William Homan Thorpe ferried recordings of chaffinches back and forth between the Teddington spectrograph and his job at Cambridge University before finally getting the funds for his own machine through a Rockefeller grant.[12] As the story goes, the pictures of finch song created in that early study launched a new phase in the objective study of birdsong.[13]

For Thorpe, the spectrograph was a nonhuman voiceprint, generating laboratory-quality evidence about animal song. It was probably during his visit to Bell Labs in the early 1950s that Thorpe saw his first complete spectrograph, the ideal tool for a birdsong laboratory.[14] Unlike musical notation and mnemonics, the spectrograph was able to translate entire sounds—pitch, timbre, and rhythm—directly into images. The tool seemed to resolve the shortcomings of musical notation, which could not represent timbre; but more important, it seemed to resolve the differences between human and bird hearing. Many birds sing very quickly in a register at the upper end of the human hearing range, so musical notations were especially hard to make even with the aid of recorded examples. Mnemonics were often even less accurate, serving as a memory aid rather than a description.

Figure 6.1 shows a comparison of mnemonics, musical notation, and a spectrogram, using three representations of the chaffinch's song from an 1896 musical transcription by Charles Witchell, a 1922 example of mnemonics by Walter Garstang, and a 1954 spectrogram by William Thorpe. Thorpe's spectrogram contains a richness of detail that is much greater than either mnemonics or musical notation. By transferring sounds whose speed and register fell outside of human sensory norms into images that fell well within them, the spectrograph provided information that was easily apprehended without special musical training, and at a much higher level of accuracy than the human ear was able to offer. It provided more melodic detail than musical notation did, and even offered a way to visually notate sound's timbre in the thickness of its lines. As a detailed visual record, the spectrogram surpassed musical notes and mnemonics with ease. The spectrograph's image was not only more accurate, it appeared to remove subjective elements of humanized, culture-specific interpretation, for the machine did away with the anthropocentric language of Western music notation and kitschy mnemonic catchwords. Thus ornithology's lore describes the spectrograph not only as the mark of a new phase in the study of birdsong, but as the origin of the *objective* study of birdsong.

Yet the way Thorpe used the new tool suggests that there is more to the notion of objective interpretation than meets the eye. There is a curious addition

Chip, chip, chip; Tell, tell, tell; Cherry-erry-erry; Tissy-choo-ee-o!

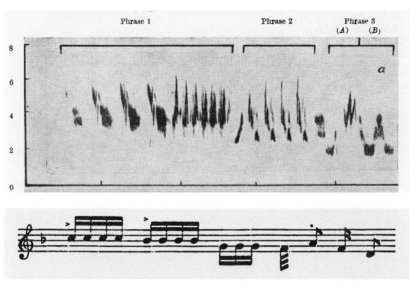

FIGURE 6.1 Three representations of chaffinch song. (*Top*) Mnemonics by Garstang, *Songs of the Birds*, 37; (*bottom*) notation by Witchell, *The Evolution of Bird-Song*, 235; (*middle*) spectrogram by Thorpe, "The Process of Song-Learning in the Chaffinch as Studied by Means of the Sound Spectrograph," 466.

to Thorpe's image, the manual subdivision of the graph into phrases 1, 2, and 3. Looking at the graph alone, the image could easily suggest up to five phrases. The picture suggests that the second half of Thorpe's phrase 1 could well be an independent phrase, and that the upbeat to phrase 3, which Thorpe cannily slips between his brackets, is another independent element. Yet both Garstang's and Witchell's renditions of the chaffinch are very clearly grouped into three elements, not four or five. Thorpe's agreement with them on this point suggests that his phrase grouping does not come from an inability to read his own graph, but from whatever he heard in the chaffinch's actual song. That elusive formal shape was something that Thorpe, Garstang, and Witchell all heard in the same way. It was the spectrograph that heard the phrase differently, not the human listeners.

So in matters of acoustics such as frequency or overtones, Thorpe trusted the spectrograph over his own ears. But when confronted by a conflict about how to interpret what counted as a phrase, it was human listening that Thorpe seems to have selected as the closer representative of the chaffinch's experience. In print, Thorpe emphasized the superiority of his new mechanical ear over "the naked ear" of a human listener: his machine could document elaborate patterns accurately, had a perfect memory, and could hear a bird's high speeds and frequencies better than a human being could.[15] These benefits were very real advantages. But Thorpe's new objective listening still required the kinds of aural judgments that had guided Hornbostel's listening techniques at the Berlin Psychological Institute three decades before. (In fact, the psychologists in Berlin had used musical phrases as one of the first test cases for their Gestalt integration of experimentation with aesthetic holism.) The difference for Thorpe was that the problem of interpretation, though still present, became invisible in the shadow of the spectrograph's apparent objectivity.

This slight disjunct between rhetoric and practice may be a symptom of some of the more pragmatic concerns that would have made the spectrograph about more than just accuracy for Thorpe. In 1950, he was in the midst of a key professional transition from being a mid-level professor at Cambridge to becoming the head of a new laboratory, Cambridge University's Ornithological Field Station at Madingley.[16] Madingley, in turn, would eventually become a department at Cambridge, the university's first program in animal behavior or, as Thorpe called it, ethology. Like Agassiz, Davenport, and Whitman, Thorpe was making a transition from professor to laboratory director in the 1950s.

Ethology, Thorpe believed, needed to carve a middle path between the specialist's laboratory and the naturalist's intuition, between objective and subjective ways of knowing animals. In the sphere of the laboratory, he saw a tradition of rigid behaviorism inspired by Morgan's canon and passed down by Russian and German physiologists who tested animal reflexes in strictly controlled experiments at the turn of the century. In the sphere of the naturalist, Thorpe saw a contrasting notion of holistic learning, explored in the early 1900s by psychologists who investigated open-ended questions about animal consciousness and learning.[17]

The spectrograph didn't just provide Thorpe with an abstractly objective ear; it helped him navigate between these worlds of physiology and psychology. The "rigid behaviorists" were epitomized for Thorpe by Jacques Loeb and John B. Watson at the University of Chicago's department of physiology.[18] For the behaviorists, objectivity was grounded in the evidence of the physiological laboratory and its carefully calibrated instruments. The foil of this approach, in Thorpe's view, was the new psychology practiced by the founders of Gestalt theory from the Berlin Psychological Institute: Hornbostel's research partner Max Wert-heimer, Kurt Koffka, and Stumpf's successor at the institute, Wolfgang Köhler.[19] Influenced by Stumpf's penchant for philosophy, Gestalt's practitioners argued that perception was a holistic experience and couldn't be successfully described by partitioning body, mind, and sensation. Gestalt's followers relied less strictly on the instruments of physiology, instead favoring human subjects whose accounts were factored into experimental results. For Thorpe, the behaviorists were too limited in their treatment of animals as machines with no thoughts or feelings; but the comparative psychologists threatened to teeter into personal anecdote, and had already been harshly criticized before the war for being overly subjective in their methods.[20] (Though he didn't say it, Thorpe also probably had Whitman and Craig in the back of his mind, who were at Chicago's zoology department while Loeb and Watson were there. Thorpe had visited Craig in 1951, but didn't seem to know quite how to fit him into this pantheon of animal behavior, an oversight later corrected by Thorpe's student Peter Marler.)[21]

By adding the spectrograph to the field station at Madingley, Thorpe hoped British ethology would emerge as a middle path between these two extremes, asking Gestalt-style questions but using physiological-style instrumentation to quantify his results. Birdsong seemed to be the ideal subject matter to fit between these poles, combining genetic inheritance and cultural learning in a single act. Prior studies of birdsong, reliant on musical notation, had failed to generate aural evidence that met the standards of quantification established in the physiological laboratory. Just as the old German and Russian reflexologists had used *Apparate* to generate knowledge, spectrograms promised to help ornithologists leapfrog from the grooves of the sound recording to more easily quantifiable visual data. Spectrograms of birdsong would leverage comparative psychology's questions about learning into the objective realms of physiology.

When Thorpe published the results of his chaffinch study in *Nature* in 1954, he argued for that nuanced middle ground between cultural and genetic inter-pretations of chaffinch songs.[22] Songs were both learned and innate, he claimed, comparing the differing elements in songs of birds who grew up with tutelage from their biological finch parents to those of orphans or babies raised by canaries or goldfinches. At the center of his research were the images of the spectrograph, which he placed side by side to compare what he called the "normal song," a standard pattern virtually identical to the prewar notion of a sonic typology, with songs learned in nonnormative situations.[23] While Thorpe valued musical listening skills, he also noted that the spectrograph's image was vital because "even the most musical find it difficult to remember minute details of song without the aid of a suitable notation; so the dangers of subjective interpretation are always serious."[24] Thorpe did not specify exactly what kind of subjective dangers awaited the ornithologist. But he did note how much easier it was to see the rapid patterns and high pitches of birdsong in the spectrograms. And its images were compelling, providing a way for readers to compare the songs of his research subjects in *Nature*'s pages without recourse to a sound recording.

Thorpe's advocacy of the spectrograph was tempered by a lifelong respect for musical knowledge. His colleague Robert Hinde once said that Thorpe's life was "a struggle to reconcile his aesthetic appreciation of music and of nature with the objective realities of entomology and traditional ethnology."[25] At Cambridge, Thorpe played Liszt and Busoni on the piano and passionately collected 78-rpm records.[26] Like older ornithologists, Thorpe used his musical skills in his work, describing birds' songs with musical notation and analyzing specifically musical features of a song such as harmony, polyphony, or antiphony.[27] Thorpe's laboratory at Madingley was in many ways what Wallace Craig might have built, had he had the resources.

Despite Thorpe's connections to prewar notions of musicality and comparative psychology, and despite his efforts to craft that middle road between Gestalt and behaviorism, he is best remembered for his contributions to "objective reality." The authors of one history of avian bioacoustics wrote in 1994 that "avian bioacoustics passed from anecdotal to objective description" with the spectrograph.[28] A decade later, Thorpe's student Peter Marler reminisced, "Until about 1950, everyone in birdsong had no choice but to work by ear. Only when the sound spectrograph became available was it possible, for the first time, to grapple objectively with the daunting variability of birdsong, and to specify its structure precisely. Almost immediately a multitude of new issues became accessible for scientific scrutiny and experimentation."[29] In the story of Thorpe's research, musical subjectivity was exchanged for the visual quantification of the spectrograph in 1950.[30] The advantages of comparative psychology and musical listening, which Thorpe consistently championed, are submerged under the broader message. The sound spectrograph brought studies of animal vocalization into the empirical laboratory for Thorpe's biographers, a process epitomized by the word used so often by Thorpe and later ornithologists:

"Objectivity."

With that word, I turn from birds to human beings and the parallel history of graphic song notation, provided by music scholar Charles Seeger and his invention of the melograph. Seeger's story begins, as he tells it, with a chance encounter with none other than Erich Moritz von Hornbostel, the musical vivisector of chapter 4. In addition to Hornbostel's work in comparative psychology, the doctor had coauthored the period's most comprehensive survey of methods of transcribing music in 1909 with his colleague Otto Abraham.[31] Included in their comprehensive article was a discussion of the merits of graphic song transcription. In the early 1930s, as Dr. Hornbostel was preparing for his never-to-be-fulfilled move to New York City, Seeger asked him what he knew about sound-writing machines. Hornbostel, in response, directed Seeger to an engineer working at Bell Laboratories.[32] The engineer (Mr. Weigl) believed the machine Seeger wanted was possible, but so expensive it wasn't worth building. Seeger writes that he dropped the matter but returned to it a decade later, when Ralph Potter publicized his own work on the spectrograph with Bell Labs. After reading about the spectrograph, Seeger enlisted the help of an engineering company in Chicago to reconstruct Bell's filtering system at a lower cost. Working with an unimpressive diagram and his knowledge of Potter's machine, Seeger was able to borrow a laboratory at Cornell University over a weekend and build a prototype for his own version of the spectrograph. That prototype, after considerable tinkering, became the basis for the model C melograph, positioned to take on the same role in music that the spectrograph was taking on in studies of birdsong.

The melograph met many of the same notational needs that the spectrograph did. Composer Béla Bartók's oft-repeated words, "the only true notations are the sound-tracks on the record itself," voiced a desire for a new level of accuracy in music studies that mechanical transcription seemed to offer.[33] Built on a guess and a limited budget, the melograph didn't successfully replicate the way the spectrograph filtered sound waves. It did, however, create a creditable graph of a single melody line, making it well suited to the comparison of songs side by side. Seeger recommended that the machine's images, or melograms, be supplemented with musical notation, as in his melogram of "An Maighdean Mhara," shown in figure 6.2, so that researchers could visualize music in two ways.[34] Music notation told you how to perform the song; the melogram was a record of how a singer had actually sung it in practice.[35]

For Seeger, the melograph promised to shift postwar music scholarship into the same kind of prestigious laboratory terrain that Thorpe had found at Cambridge. The clerical work of song collecting was well underway; real analysis of those collections, Seeger felt, was now possible. This was the final, incomplete half of "two great steps" in music research, he wrote; "First . . . *sound-recording in*

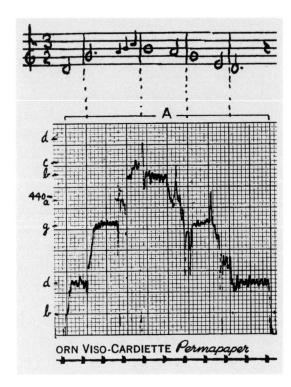

FIGURE 6.2 Excerpt of a melogram by Charles Seeger, showing the opening phrase of "An Maighdean Mhara" as sung by Kitty Gallagher. Seeger, "Toward a Universal Music Sound-Writing for Musicology." Copyright International Council for Traditional Music (ICTM).

*the field*, and second, . . . *sound-writing in the laboratory*."[36] But actually launching a music laboratory was difficult. Stumpf and Hornbostel had made their case for the Phonogramm-Archiv with the help of the preexisting lab at the Berlin Psychological Institute, and with considerable funds from Stumpf's own pocket. Seeger had no such umbrella of support. In 1953, however, he was appointed to the music department at the University of California, Los Angeles, joining Mantle Hood in the school's developing program in ethnomusicology. Though UCLA's music program never transformed into the kind of lab Thorpe had at Madingley, it gave Seeger access to institutional funds, space that allowed him to develop the melograph, and a host of students eager to use the new machine. By 1968, the university had fronted over 85,000 dollars to develop the instrument into a viable apparatus, and Seeger began to market the upgraded machine as the "model C."[37]

Though Seeger never explicitly located the melograph in the middle path between behaviorism and Gestalt that Thorpe described with such fluency, the machine was nevertheless loosely connected to the same lineage of comparative psychology. Seeger credited Hornbostel with his introduction to Bell Labs; and most of the melograph's later adherents were once-removed acolytes of the Berlin Psychological Institute's students. The head of the new ethnomusicology program at UCLA, Mantle Hood, had studied at the Berlin Psychological Institute; and so had George List, who later became curator of a collection of Hornbostel's recordings at the University of Indiana.[38] (With Seeger's help, List used the melograph to reconceive Benjamin Ives Gilman's "hortus siccus" of Hopi songs from the early 1900s, rewriting Gilman's graphic notation as a mechanical transcription in the 1960s.)[39]

As with the spectrograph, the rhetoric of the melograph was the language of objective and subjective knowledge framed by the implicit association of mechanical transcription with objective research, and musical listening with interpretive subjectivity. From the very beginning of his enterprise, Seeger adopted this language, promising that the melograph would provide "an objective record" of song.[40] The tool was "descriptive and objective" by its nature, in contrast to the prescriptive, subjective practice of aural transcription, and therefore it offered a form of notation "that can be written and read with maximum objectivity."[41] Musical notations, in contrast, were "prescriptive and subjective" and led to a "thoroughly unscientific" notation, "a riot of subjectivity."[42] It was a one-man echo of the story of the spectrograph, reinforcing the postwar apotheosis of graphic transcription technology.

## SUBJECTIVE LISTENING

Etymologist C. T. Onions wrote in the 1960s that objectivity comes from the Latin *ob jacere*, to throw a thing (*jacere*) before oneself (*ob*).[43] More recently, historians Lorraine Daston and Peter Galison wrote of mechanical objectivity, in which photographic images threw unmediated reality before the cautious eyes of scientific atlas makers at the start of the twentieth century.[44] Half a century before Thorpe and Seeger, when those photographs charmed science's atlas makers, song collectors were more interested in the practical challenges of representing new and peculiar sounds than in the opposition of subjective and objective knowledge. Unworried by their subjective selves, they merely called innovative transcriptions "baffling,"[45] complaining that the visual contortions required to represent non-Western sounds made a "regrettably disturbing impression to the eye" when transfixed in musical notation.[46] Yet these exotic songs did inspire a handful of collectors to yearn for a mechanical sort of objectivity, something that was similar to, but not quite the same as, the objectivity of Daston's atlas makers, for it was shaped by the unique needs of musical research. This yearning for mechanically produced images was especially strong in the song laboratories of the 1920s and '30s that operated within the realms of experimental psychology. There, ethnographers and psychologists spoke about "objective methods" of describing songs, "objective measurements" of music, and songs that would themselves become "objects."[47] What exactly did mechanical pictures throw in front of these researchers that musical notes did not provide?

Experts in song generally used the word "objective" to mean one of two kinds of images. The first, rather counterintuitively, consisted of handmade graphs. This imagery was not about detail but about form, the picture of musical vivisection described in chapter 4 as life's unpicturables, the general laws of musical behavior that required broad generalization rather than accurate details. Images of birdsong, exotic music from Asia, and indigenous tribal songs were traced with painstaking care as researchers compared what they heard on a recording to pitch pipes and a stopwatch, plotting the results on graphing paper. Those results were then used as the scientific basis of laboratory deductions like Hornbostel's that led to smoothed-out, less detailed images that revealed music's typologies.[48] These graphs were the picture of difference, the image of music's prewar typologies. The handmade nature of such images, which imitated the look of a sound-writing machine without actually using one, suggest that it wasn't possible to make the same graphs using a mechanical device. Yet there were at least a handful of machines before World War II that created true graphic inscriptions of sound: the phonautograph and similar machines, and the rolls used in player pianos and music boxes, which were also considered graphic notation.[49] It was as if the graph itself was more important than whether it was produced mechanically.

Indeed, there were quite a few other options for the mechanical transcription of sound, thanks to a family of machines that photographed images of sound using light. Rudolph Koenig, the German *Apparat* builder who added the kymograph to the original phonautograph, designed a manometric flame apparatus in the nineteenth century that served as the basis for twentieth-century devices that captured sound in the flickering patterns of vibrating mirrors.[50] These images, too, were a second kind of valued "objective" rendering of music, and often the same individual would experiment with both handmade graphs and photographic images in the quest for objectivity.[51]

With this association between the (photo) graph, objectivity, and sound in mind, turn back with me for a moment to Thorpe and his middle path between behaviorism and Gestalt psychology. Behaviorism was founded in physiology—a classic inspiration of the school was Pavlov's famous experiment where a dog salivates at the ringing of a bell. For the behaviorists, the experiment wasn't merely interesting because of the bell and the dog, but because of the tubing apparatus Pavlov had inserted into the dog's salivary glands, which he could attach to a vial that would measure exactly how much the dog salivated, enabling Pavlov to quantify desire in saliva units. Evidence, for strict behaviorism, was the ingenious measure of quantity enabled in a very literal way by prior experiments in vivisection. The Gestalt psychologists whom Thorpe imagined as physiology's foil, in contrast, sought a more culturally based standard of evidence, inspired by physiology without being defined by it.

Hornbostel's attempts to measure musical form in the Phonogramm-Archiv were in keeping with this experimental style, in which experimenters might ask what the difference was between hearing individual notes and hearing a complete melody.[52] When Gestalt psychology was criticized in the 1920s and '30s, it was for straying from the physiological standards practiced by experimentalists such as Pavlov.[53] For its detractors, Gestalt methods were vague and deductive, and "completely confused" subjective and objective precepts.[54] Encoded in this language was the understanding that "objective precepts" meant, on some level, physiological instrumentation. Instead of Gestalt inquiries, which focused on aesthetics, emotion, and learning, critics were arguing that the standard of scientific knowledge in studies of music or culture should be the standard of physiology—the intelligent reflection that overcame the researcher's emotions, epitomized, as has been shown in chapter 4, by the insertion of the kymograph into the quivering body of an animal.

Without explicitly naming the Gestalt school, song specialists in the 1930s began to worry that they needed a machine in order to create truly objective evidence. This rhetoric was most noticeable at the psychology program at the University of Iowa, which turned in the 1930s to comparative song research in which "the objective method" was virtually synonymous with the graphic image.[55] The lab, run by Charles Seashore, had begun a large-scale study of musical eugenics in collaboration with Charles Davenport at Cold Spring Harbor in the 1920s. When the results of that work proved almost impossible to quantify, Seashore and his students began to turn instead to a variety of mechanical transcription devices, including a kymograph attached to a phonograph,[56] before settling on a photographic method called "phonophotography." Seashore's students used their photographic methods to churn out pictures of birdsong, African American spirituals, and Native American speech throughout the 1930s. Why, they asked, would you use the "subjective method" of human listening when it was possible to photograph objective images of sound?[57] Similar questions followed comparative studies of animal voices. Just a few years after the invention of phonophotography in Iowa, a similar process using movie film was developed by an ornithologist at Cornell, who wrote pointedly, "We cannot, try as we will, hear objectively; it is impossible to separate the hearing apparatus from the thinking mechanism—the ear, from the brain. Hearing is a decidedly subjective function."[58]

At stake was the ethics of knowledge defined in the physiological laboratories of the nineteenth century, where *Apparate* mediated the exchange of life for knowledge. Though researchers like Hornbostel had transferred the spirit of that exchange to aesthetic inquiry, Gestalt's critics insinuated that the results lacked the kind of material evidence that could be obtained through literal investigations of living bodies.[59] Although it is not clear whether Thorpe's adoption of the spectrograph was an explicit response to those critiques, he framed his work more broadly as a response to the dynamic between behaviorism and Gestalt. And when Seeger described his first attempts at the melograph, he credited the students of Seashore's laboratory as his inspiration.[60]

Perhaps it is no coincidence that when Thorpe and Seeger applied the rhetoric of empirical laboratories to their work, they adopted not just the language of prewar experimental psychologists, but the mechanism that most closely resembled the nineteenth-century physiologist's kymograph. Of all of the devices that offered methods of mechanically transcribing sound, only a handful resorted to the kymograph's model of writing vibrations transmitted directly with a stylus onto a rotating cylinder, rather than photographing sound. The phonautograph was one, and its form was adopted by several of the later inventors at Bell Labs, including the prototypes that produced the spectrograph and the melograph. Both machines transformed the electrical variations produced by recordings into a quivering stylus whose point inscribed sound onto a rotating cylinder wrapped with paper, the direct descendant of the kymograph (figure 6.3).

## PARTING OF WAYS

The main points of my story thus far are as follows: During the 1950s, William Homan Thorpe and Charles Seeger independently inspired mythologies of objective mechanical transcriptions of sound. Closer inspection of their stories reveals preexisting machines very like theirs, and preexisting qualms about the

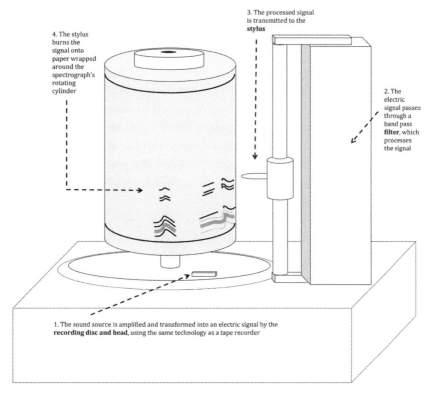

4. The stylus burns the signal onto paper wrapped around the spectrograph's rotating cylinder

3. The processed signal is transmitted to the **stylus**

2. The electric signal passes through a band pass **filter**, which processes the signal

1. The sound source is amplified and transformed into an electric signal by the **recording disc and head**, using the same technology as a tape recorder

FIGURE 6.3  Diagram of an analog sound spectrograph.

subjectivity of the Gestalt school's behavioral research. Those qualms, in turn, raise the question of what was at stake in the mechanical transcription of sound.

The mythographer Joseph Campbell wrote in 1949 that the story of a hero is a story of self-achieved submission. "But submission to what?" asked Campbell.[61] The heroes of mechanical transcription in 1950 accomplished multiple ends. One of those ends was the representation of song in a mechanical format that, when operating correctly, provided a one-to-one relationship between the pitch and timing stored in a source recording and a graphic image. Another was the self-achieved submission to objectivity, as it was defined through the graphic traces of the kymograph and the *Apparate* of the physiological laboratory.

For ornithologists, that meant submission to the spectrograph. Historians situate the beginning of "objective" animal vocalization research with Thorpe's lab at Madingley and the new machine he used there.[62] Vision, not hearing, dominates these stories; as American ornithologist Donald Kroodsma explained in 2004, "I have well-trained eyes, and it is with my eyes that I hear."[63]

Yet this submission came with a price. Researchers at Madingley had to fiddle with the spectrograph's speed to create a legible image of chaffinch song. The machine was calibrated for human speech, so researchers working with birdsong had to slow down or speed up recordings to increase or decrease the size of the finished spectrogram image.[64] The editor of *Ibis*, Britain's leading ornithological journal, complained that even the best images were so faint they couldn't be reproduced in print. Thorpe's student Marler would have to trace them onto paper, using calligraphic lines he had seen in the graceful brushstrokes of Japanese ink paintings.[65]

Another idiosyncrasy was the paper roll attached to the cylinder. On the melograph, Seeger had thought to attach the spool of paper in a continuous roll, so the image could record an entire melody. But the spectrograph's written image was limited by the diameter of its rotating cylinder, for its makers at Bell had designed it to carry only as much paper as would wrap around the drum. As a result, it could graph only about two to four seconds of sound at a time.[66] Older studies of songbirds had focused on long and complex songs: the hour-long concerts of the mockingbird, the brown thrasher, or the European nightingale. Even Craig's wood pewee study, which was based on a bird with very short phrases, was focused on the long sequence of phrases before dawn. Submitting to the spectrograph meant accepting that these birds were impossible research subjects.

Accepting those limitations, researchers began to focus on birds with short, highly repetitive songs, especially if they were easily bred in captivity, so that a single two-to-four-second sample could serve as a potentially representative example of species behavior.[67] Species such as the chaffinch, which became Thorpe's research subject, and the white-crowned sparrow, which became Marler's specialty, fit this model and dominated new research. Even after digital technology made it possible for computerized spectrographic programs to analyze longer recordings in the 1970s, the two-to-four-second sound clip remained the norm in laboratory studies of avian vocalization, perhaps partly because it reproduces well on paper, running about the width of a page.[68]

Music, in some sense, was the price of this submission. Ornithologists did not lose their interest in attentive listening to longer songs, but this was harder to talk about as part of their work. Kroodsma, who claimed to hear with his eyes, admitted later in a less formal setting that his ears did do vital, but suspiciously imaginative work: "By truly listening, I can know a bit of what the singer is 'thinking,' and I'm a little bit closer to understanding what it is like to be the bird. And that, after all, is my ultimate goal, to try to imagine what it is like to be the bird in that moment."[69] Thorpe's student Marler likewise admitted that he sometimes wondered about the music of birds, but made clear that this kind of listening was not real science.[70] These confessions pose musical listening as a breach of scientific conduct, made irresistible by the pleasure of imagining avian subjectivity, "what it is like to be the bird."

The melograph, too, demanded heroic submissions. Seeger was constantly tinkering with the machine.[71] Like Marler, the students at UCLA had to trace melograms to make them publishable, decorating them with beautifully written musical notations instead of Japanese brushstrokes. The stylus moved more slowly than the electrical signal. Seeger's students sometimes noticed hearing sounds that the machine didn't seem to transmit. And the machine rejected any sounds it deemed too complicated, including guttural sounds and, apparently, the Hungarian language.[72] Adapting meant studying only melodies: no harmonies or duets, no drums, no guttural sounds, and no throat singing. No Hungarian.

At first, Seeger and his colleagues followed the hero myth, submitting to the machine's needs in their quest for objectivity as ornithologists had with the spectrograph. But by the early 1970s, their story began to diverge sharply from that of ornithology. George List, the student of George Herzog who had used the machine throughout the 1960s, decided to compare transcriptions made by ear to those made by the melograph. After setting several graduate students to the task of making musical notations and comparing the results to corresponding melograms, List drew the "inescapable" conclusion that melograms were no more accurate than human listeners, and in some cases were less so.[73] The machine's objectivity was, in fact, a fraud.

List's study was followed by an even starker critique of the machine by Nazir Jairazbhoy, one of the students who had studied with Hood and Seeger at UCLA. Rather than test the machine's objectivity, Jairazbhoy questioned the entire laboratory tradition, arguing that the melograph showed ethnomusicologists that "the terms 'objective' and 'true' surely need qualification."[74] Jairazbhoy argued that it was not the electronic device that created musical information, but a human listener who had personal and political stakes in the creation of musical categories.[75] The question was not whether the melograph was really objective, but more fundamentally, what objectivity was, and why it was valued. By the end of the 1970s, List's and Jairazbhoy's debates had relegated the melograph to an entirely different place than that of the sound spectrograph. Though the melograph did not entirely disappear, it was excised from the field's hero mythology, becoming largely obsolete by the end of the century.[76]

It was a remarkable shift. For the first half of the twentieth century, animal voices and human songs had been united by the common tradition of experimental and comparative psychology. But by the 1970s, the two had diverged, and sharply.[77] Studies of birdsong followed the path of the professional research laboratory, becoming firmly situated within an accepted notion of science in which knowledge gathered in the field was transformed into data in the laboratory by a process of graphic transcription. Those who studied human song, however, rejected the laboratory model and with it, the value of the objective graphic image. By the 1970s ethnographers no longer wrote of their ties to the kymograph and the laboratory animal. Yet the mythology of transcription is easily summoned to exorcise the ghosts of the psychological laboratory from ethnomusicology: "Transcription . . . eliminated the life-world, and transformed what was left (sound) into a representation that could be analyzed systematically and then compared with other transcriptions so as to generate and test hypotheses concerning music's origin and evolution. Today it is not transcription but fieldwork that constitutes ethnomusicology."[78]

Today, ethnomusicology and animal vocalization research are unrelated disciplines. The first engages in the study of human music-making; the second investigates the sonic artifacts of nonhuman species. The former is about a human, musical subject; the latter is about an animal, nonmusical object. The story of mechanical transcription after World War II is the story of this divide, and of the radically different traditions of evidence creation that resulted from it. Animal musicality, though a question of interest to postwar ethologists, demanded a subjective act of imagination that had become taboo in the laboratory. Yet that same act of imagination became the core of postwar ethnomusicology: the informed imagining-of-the-other that was the task of the ethnomusicological hero.

Amid this bifurcation of knowledge, singing birds, who had been the analogue of both race and human musicianship for over seventy years of Western science, were divided from the cultural imagination. In the aftermath of that division, musical evidence is made of a fundamentally different material than the material of biological evidence: the stuff of Gestalt imagination and musicality is not the stuff of graphic images and genetic neuroscience. It is not ornithological veracity that captures my attention with this divide; that was a problem of rendering nonhuman song accurately. Rather, I am intrigued by the curious opposition of subjective and objective knowledge that came to dominate postwar transcription.

That rhetoric is not quite the same as Descartes's problem of body and soul, but a dilemma about the role of one's personal experience in the acquisition of modern knowledge, and, more fundamentally, about the relationship between the human researcher and those he (and sometimes she) studied. Postwar subjectivity was, quite simply, the problem of how to deal with difference, with otherness, that was felt within oneself. In the postwar decades, the stakes of this dilemma were starkly framed by the memories of scientific racism, the lived reality of racial segregation, and the hope of resolution proffered by genetics that was discussed in chapter 5.

With this thought in mind, I wish to turn back farther, to remember the problem of difference that preceded postwar notions of subjectivity in the collection, classification, and analysis of songs. Thorpe and Seeger's joined story points to a broader set of questions about the lived experience of difference that found expression in the collection of sonic specimens discussed in chapter 3. In the years after World War II, that experience was reconfigured by the construction of postwar human identity around a division between culture and genetics. The evolutionary synthesists framed difference as a genetic problem; ethnographers, anthropologists, and musicologists framed difference as an ethical problem. Between genetics and ethics, the stage was set for the disparate ways of knowing animal and human song that emerged in Thorpe's and Seeger's work during the 1950s. The result was a divergence in disciplinary notions of evidence, one for studies of human beings, and another for studies of singing birds.

Who could bring those two halves back together again? How? For in these two worlds are two disparate models of knowledge and knowledge-making, framed by a century-old conundrum about the experience and ethics of being different. That question leads not forward but backward, to the song collectors of the prewar era.

## CODA

I end this story with a coda, a might-have-been, which suggests the many openings that remained within the postwar scientific purview of the subjective ear. Wallace Craig's biographers Mills and Kalikow write that when Thorpe delivered a guest lecture at Harvard in 1952, he was astonished to hear that Wallace Craig was in the audience. "I assumed him to be dead!" Thorpe proclaimed.[79] Craig, far from being dead, was still pursuing his own ideas when Thorpe and Seeger began their quest for objectivity. From a state of increasing poverty and isolation, he had received a grant in 1944 from the American Philosophical Society in Philadelphia to pursue a new project, which he titled "The Subjective Space System."[80] The project suggested that subjectivity was a necessary and positive element of perception, and included a chapter on aural subjectivity. Craig asked experimental psychologists to become "naïve . . . [and] forget all that you have learned about the human body as an object."[81] Subjectivity, he argued, was important to human researchers.

Objective and subjective knowledge had occupied Craig since his wood pewee manuscript. In that book he had defined "objectivity" in opposition to existing norms. Instead of advocating for the empiricism of the physiological laboratory, Craig suggested something so bizarre that, buried in the middle of his prose, it seems to have passed over the heads of his reviewers: that objectivity was best described by the aesthetic taste of woodpeckers. Woodpeckers, Craig explained, made careful choices about the sounds they liked best. Though nature should have programmed them to prefer hollow wood, they sought out iron roofing to produce loud clanging sounds. Craig saw this as a conscious aesthetic choice, an "intelligent esthetic choice" that he seems to have loosely compared to the intelligent reflection of the more traditional scientist.[82] Ignoring common usage of the word, Craig argued that "objectivity" ought to be defined in terms of this intelligent aesthetics. "The sound of drumming is thus an object and *objectivity* is one of the chief characteristics which distinguish a work of art from a mere expression of emotion."[83] This was an objectivity that distinguished the artist, not the scientist, from lesser creators.

Craig must have started his new study of subjectivity before his book on the wood pewee was completed, for he shared the first drafts of his new project just a few months after *The Song of the Wood Pewee* was published. The manuscript was sent for review to Wolfgang Köhler, the one-time director of the Berlin Psychological Institute before the war. After denouncing the dismissal of Jewish scientists in the 1930s, Köhler had fled Germany, and now worked at Swarthmore College. He was startled by the manuscript's original thoughts, and wrote positively about Craig's work, but expressed frustration that its author seemed completely unfamiliar with recent psychological literature from Germany. Craig didn't seem to know much about Gestalt writings, or about their critics. He was unprepared to make the kind of intervention Thorpe had made, responding to the critics of subjective science. The whole thing, Köhler suspected, would have to be rewritten.[84] Craig was asked to remove the word "subjective" from the paper.[85]

Craig's manuscript, which was never published, has been lost. But it points toward the human beings for whom subjectivity remained a necessary question in the sciences as well as in studies of culture: the scientists who supported his grant with the American Philosophical Society, and even Köhler, who hoped for a revised publication. What they saw in Craig's manuscript seems to have some common ground with ornithologists' persistent desire to imagine avian subjectivity, and with ethnomusicology's postwar commitment to a subjective ethics of knowledge. In the next chapter, I explore those ruptures with the myth of objectivity by looking toward the imagined sounds of a postwar, postmodern, posthuman Garden of Eden, turning from the traces of the kymograph to the joys and troubles of its heritage in the postmodern soundscape. There the sonic specimen is revived in the sound recordings and music-making that represent postmodern identity through the pedagogy of natural identification, taught in the instruction books of listening creatively to nature's culture.

SEVEN

# The Rose Garden

O bird of the morning,
Learn love from the moth
Because it burnt, lost its life, and found no voice.

*Sa'di,* The Gulistan

The reception history of birdsong has yet to be written, but it will surely dwell on the audio field guide. These guides train listeners to identify birds by ear, drawing on the typological collection and classification that made sound a part of natural history in the early twentieth century. Such guides have been issued since the 1930s in the form of phonograph records, tapes, and compact discs. Today they are increasingly popular, taking the form of digital files embedded in apps, websites, or books where sound is accessed at the tap of a finger.[1] Such guides connect the project of hearing species difference with the goal of hearing spatial difference, in the meadows, forests, and mountains of the past—"the field" of the field guide. To do this, guides increasingly rely on creative formats that verge on composition in order to depict nonhuman geographies in sound. In the process, their pretext of objectivity gives way to complicated patterns of ownership, nostalgia, and discovery. While audio guides purport to educate listeners about natural identity, they also model important relationships between human listeners, singing birds, and imagined spaces of nature. In this chapter, I turn to three of these imagined spaces to ask how listening to birds models interspecies ethics in a postwar, postmodern, posthuman world.

I conclude my book with three soundscapes: Olivier Messiaen's *Catalogue d'oiseaux* (1956–1958), Steven Feld's *Rainforest Soundwalks* (2001), and Miyoko Chu's *Birdscapes* (2008). Chu's book, Feld's recording, and Messiaen's work for piano are, in very different ways, soundscapes that merge aspects of a field guide with a different, more aestheticized tradition. The terminology of the soundscape was coined in the 1970s by Murray Schafer, about fifteen years after Messiaen's *Catalogue*.[2] Schafer's thinking was intended to exist, like the song of a bird, in a cultural in-between. His involvement in soundscape studies and acoustic ecology was a highly aesthetic venture that, like the examples I explore here by Messiaen, Feld, and Chu, invoked an exotic habitat he called the Soniferous Garden, a "feast for all the senses" in which the sounds of nature reigned.[3] Schafer's Edenic terminology of "soundscape" has entered mainstream music literature, although it retains less of its original flavor.[4] The term soundscape is also used as a tool in ecological circles to defend the idea that local animal habitats are disrupted by the introduction of industrial sounds, an idea promoted by Schafer under the phrase "acoustic ecology."[5]

Acoustic ecology and the soundscape are reminders of the inherent conundrum of the objective observer, who attempts to record and identify nonhuman sound without registering a human effect. Creators record, crouching in the brush with a specialized microphone; they choose which recordings to retain and which to discard, and they choose which parts of those recordings will best teach the sound of a species. The users of these creations often encounter a notable contrast between their own lived experiences and the pristine forests, prairies, and tundra of the audio guide. These encounters beg the question of what the end goal of species identification is, and what it should be, for users of the modern audio guide. Field guides to the nonhuman therefore present us with a riddle that has been lurking in musical evolutionism since its inception: What is the life of the objective listener, and who is this individual? And where is paradise in that world where nature is divided from the privileged class of humankind?

Edenic soundscapes are the subject of this chapter, which explores the interplay between postmodern human identity, ethical relationships, and imagined Edens. In Judeo-Christian tradition, the sacred nature of the Garden of Paradise is located in the species-specific problems of human sin, guilt, and sanctity. Humans in this garden embody an ethics grounded in uniqueness that is itself particular. Centuries of iconography locate Adam and Eve in already-unique white and Western bodies. These already-unique bodies are further distanced from animals by their ethical uniqueness: humans are sinful, guilty, or sacred; animals are not. And here I come down the path that leads from Cold Spring Harbor and Cambridge ornithology to the garden of this chapter: postmodern human identity, which frames human singularity around the boundary between genetics and human culture, replicates the sacred foundation of the Judeo-Christian Garden.

The three soniferous gardens at the center of my chapter are informed in various ways by key parallels between postmodern human identity and the Judeo-Christian Garden. By listening along their garden paths, I hope to give context to the broader problems that surround the search for Eden in the present day. In the postwar, postmodern, posthuman economy, privileged human desires are profoundly more achievable than the desires and needs of those who are classed as different. Environmentalists, musicians, scientists, social justice activists, and animal rights activists have concerned themselves with this problem. Yet the ethics that highlight human responsibility in the world replicate the conditions of human uniqueness by framing human beings as uniquely sinful or saintly, guilty or innocent. As I turn to three soniferous gardens from the postwar, postmodern era, I seek to locate the problem of paradise within this contemporary context. I am not interested in evaluating or devaluing the examples I provide in this chapter. Instead, these gardens help me explore how our Edens are informed by the histories of difference within which we live, and allow me to ask how our capacity for imagination—for identifying *with* others, instead of identifying them—offers alternative spaces for ethics and new ways to imagine paradise. With that thought, I offer this chapter as a final meditation on the relationships between ethical knowledge, the study of song, and the construction of space through human/animal difference.

## FIELD GUIDES

The three cases around which this chapter turns are framed by a broader history of field guides dedicated to birds and birdsong.[6] Both audio and visual guides to birds follow the general trajectory of this book, building on the traditions of collecting and classifying song developed in the first half of the twentieth century and discussed in chapters 2 and 3. Guides to birdsong from the late nineteenth and early twentieth centuries were books of musical notation—the song sparrow who opens this book singing *Rigoletto* comes from such a guide. Guides appearing in the 1940s and later, in contrast, used sound recordings. Both old and new audio guides, however, were shaped by the visual language of the natural history museum, which was the inspiration for visual catalogues of specimens and, later, the images that became the standard imagery of picture-based field guides.

Historians of science have long noted the dominance of visual imagery in field guides.[7] Most guides draw on the visual traditions established by natural history specimens to teach observers systematic methods of viewing as a way to know and recognize animals, using images that position viewers in the role of the naturalist comparing specimens.[8] This tradition echoes aspects of the historical transition from collecting in the field to organizing specimens in the museum, and roughly parallels the chronological transition of audio guides from collections of musical notation to more professionalized guides based on sound recordings. I turn for a moment, therefore, to the interconnected pedagogies of hearing and seeing birds in twentieth-century field guides.

In the first part of this history, early visual guides presented the viewer's role as that of a literal collector of specimens, sometimes even including notes about a bird's flavor or challenges in hunting it.[9] These guides were dense with visual details about birds' habitats, prey, and the configuration of their plumage. American naturalist and illustrator Louis Agassiz Fuertes worried so much over the coloring in his illustrations that he tried to use freshly killed birds for models, so that their colors would be closer to those of a living bird.[10]

Guides to birdsong at this time, based on the musical notations of song collectors, were less obviously defined by the culture of hunting represented in visual guides. They favored the growing community of amateur naturalists who were more interested in live behavior than in hunting. Like visual guides, however, they emphasized the close examination of a single bird, providing multiple musical examples for each species and describing song in considerable detail. A case in point is Ferdinand Schuyler Mathews's description of the hermit thrush mentioned at the end of chapter 1. Preceding Mathews's guide by a decade, Simeon Pease Cheney's *Wood Notes Wild: Notations of Bird Music* was another example of this approach, with a beautiful extended description of the screech owl's song. Here, too, the listener was expected to pursue close encounters with a single bird: not only was the owl an intelligent and beautiful songster, wrote Cheney, it was also delicious.[11]

By the 1930s, the problem of knowing what to look (and listen) for when seeking natural identity had shifted more definitively to the field, favoring the new generation of birdwatchers who used binoculars and often captured no more than a glimpse of a bird. Just as proponents of the evolutionary synthesis were beginning to question the role of typology in genetic speciation, visual typologies were becoming popularized as a pedagogical tool for teaching amateur species identification. This new approach to seeing, discussed at length by historians John Law and Michael Lynch, was epitomized by the style of illustration associated with Roger Tory Peterson's guides in the United States, called the "Peterson system."[12] Developed by Peterson and launched by his publisher Houghton Mifflin in

1934, the Peterson system used simplified images with "field marks" that summed up the species' morphological type in characteristic, easily located features that a naturalist might use to quickly identify a bird in the field. The illustrations were blocky, simple images accompanied by arrows pointing out field marks: a wren's upturned tail, the red breast of a robin, or a cardinal's pointed crest. In keeping with the systematic practices of specimen typing, viewers were taught to look for these features in a definite order: first, size and silhouette; then, striking aspects of the bird's behavior; and finally, visible patterns on the breast, tail, or rump, around the eyes, or on the wings.[13]

Figure 7.1 shows an older illustration of the American goldeneye by Louis Agassiz Fuertes (named in honor of Alexander Agassiz's father Louis), alongside a more recent Peterson illustration. Fuertes's image emphasizes habitat and contextual details, while Peterson's emphasizes relative size, silhouette, and field marks. By 1947 Peterson's publisher Houghton Mifflin felt sufficiently assured of the system's success to print a new guide with thirty-six expensive color plates, far more than the meager allotment of four given to the previous edition.[14] In the 1950s and '60s, Peterson guides began to be marketed across the globe as the standard in field guides, translated into foreign languages that included French, German, Dutch, Italian, Japanese, and Icelandic.

Audio guides developed in parallel with this tradition. Modern recorded guides appear to trace their origins to a single source, the 1942 album *American Bird Songs*. The first commercially recorded guide to birdsong, *American Bird Songs* was a set of six 78-rpm records, published with support from Cornell University.[15] Made by two ornithologists at Cornell and an independently wealthy graduate of the program, the recordings were collected with the help of a special "sound truck" built in the 1930s for the express purpose of chasing birdsong in the field.[16] Originally, the recordings were intended to fill the same space as older visual guides, acting as the sonic analogue of natural history specimens for a laboratory collection at Cornell.[17] But they were also made into *American Bird Songs*.

The sounds of *American Bird Songs* set a standard for later guides that echoed the precedent of the Peterson guides. The birds in the six-record set were carefully edited to create short segments of about twenty to thirty seconds, each introduced by name. These examples included several repetitions of a short phrase of a single style meant to be representative of the species, like the field marks of the Peterson guides. Later guides issued on records, cassette tapes, compact discs, and mp3 files have followed suit. Most include around thirty seconds of song per species, in which the bird repeats a strophe, a short segment of song

FIGURE 7.1 Old and new illustrative styles: *Left*, Louis Agassiz Fuertes illustration of American goldeneye; *right*, Roger Tory Peterson illustration of American goldeneye. Eaton and Fuertes, *Birds of New York*, plate 18; from *A Field Guide to the Birds of Eastern and Central North America*, 5th ed. © 2002 by Marital Trust B u/a Roger Tory Peterson, 61. Reprinted by permission of Houghton Mifflin Harcourt Publishing Company. All Rights Reserved.

separated by pauses; coincidentally, the strophe usually falls into the two-to-four-second listening frame of the spectrograph. The aesthetic of these sound clips aims for the removal of sonic "clutter" equivalent to the details excised by the Peterson guides. Ambient sound, other species, and any sounds produced by the recording equipment or its users are edited out to produce a clean, pure aesthetic (comparable to the aesthetic discussed in chapter 6 in the context of the spectrograph). Some of these guides even borrow the visual taxonomy of picture guides, categorizing the oriole with the blackbird because both are yellow and black, or the cardinal with the scarlet tanager because both are red, despite the considerable differences in their voices.[18] Perhaps most surprising, some of the original recordings from the Cornell sound truck were used in later audio guides produced by Peterson, and can still be heard today on contemporary websites and iPhone apps like Merlin, iBird, and Peterson Birds.[19]

The aural aesthetic of these recorded guides is sometimes, as Bruyninckx puts it, sterile.[20] Reviewers generally had positive things to say about *American Bird Song* in the 1940s.[21] But they also noticed that the short examples were not always enough to learn a bird's song. As one reviewer noted, "a beginner could hardly learn to identify" common birds like the robin and mockingbird from the records, because the recorded examples were not sufficiently representative of their songs.[22] Edited for simplicity and brevity, such guides preserve and promote a particular model of sonic identity, framing the singing bird as an object of collection and classification drawn from the forgotten halls of the natural history museum. But they struggle to represent the kind of musical complexity that dominated the more fanciful guides filled with musical notation in the early twentieth century.[23]

Perhaps most curiously, the new guides often represent birds in a world of carefully crafted geography in which the postmodern human, unable to unify culture and nature, is absent. *American Bird Songs* introduced listeners to the fields, the north woods, and the prairie. But it also included birds that mingled with human life, describing game birds and birds of the garden. Later guides tend to blend such places into a new paradigm of natural space where science is packaged in fantastic geographies, where listeners encounter birds in pure, nonhuman spaces in which the sounds of the recordist, of other animals, of plants, of wind, and of traffic have all been removed. We are told of coastal mudflats and sandbars adjoining clear, open ocean surf that seem to be free of human interference; of forests thick with tree trunks and branches clustered with foliage forming a rich canopy under which birds flit, alone.[24] One guide describes in hopeful terms the European blackbird as a shy woodland resident, as if its abundant presence in crowded cities is unreal.[25] Another writes optimistically of the brown-headed cowbird as living in "open woodlands, woodland edges, agricultural areas, brushy thickets, pastures, and even suburbs" as if it is impolite to point out that most Americans meet the bird in parking lots.[26] These are avian habitats that hover on the cusp of nonexistence, hopeful landscapes where life's inequalities can be forgotten, reimagined, or temporarily set aside. Here are the places in which the postmodern human hears a nature that is forever untouched by human cultural life, where biological identity is at last fully sundered from the marks of culture.

## THE GARDEN OF PARADISE

At the threshold of this geography is the sonic representation of an untouched landscape, the illusion of a Garden of Eden. Gardens have been, from early times, places for the natural and unnatural to meet, as they do in the artificial geographies of field guides and in the many recordings that might be called soundscapes. The *Gulistan* of Sa'di's poem, which opens this chapter, was one such place of intersection, for the title means "rose garden," a garden from which the mystic, like the voiceless moth in his verse, was unable to bring back either words or sweet-scented roses. Only the poet, not the singing bird, was able to tell listeners of the presence of that garden. Recollections of such gardens seem to appear in the virgin forests, the unfarmed grassy plains, and the sparkling mountain landscapes that are the mainstay of habitat descriptions in contemporary field guides. At the edges of these pristine landscapes lurk occasional reminders that such places are rarely found in the world today, in which our gardens are not virtual metaphors untouched by the eyes of humans, but rapidly changing environments. Three years before publishing *American Bird Songs*, Arthur Allen, an ornithologist at Cornell, wrote a different kind of guide, in which birds told readers of lives beset by habitat loss and overhunting. "Miles of tundra that for years were silent, once more echo our whistles," he had the golden plover say, as it voiced its relief that its species was not extinct.[27] Years later the Peterson guide was edited to explain that many of the birds represented in its lovely pictures "have diminished alarmingly or have dropped out of parts of their range due to environmental changes."[28]

The destruction of these once-real habitats brings to mind the sense in which the original Garden of Paradise was not always a metaphor, but a source of genuine cartological debate with its own geography.[29] It was believed in medieval Europe to be at the top of or behind a tall mountain, often located in extreme regions or near the equator, in the Orient, to the east of Western Europe. Surrounding the garden or originating with it was a body of water, be it stream or lake, which was the source of the Nile. Jean Delumeau tells us the medievals also believed it to be, at various times, the source of the Tigris, the Euphrates, and the Danube.[30] Often the garden was an island in the center of a pool hidden at these heights of the world.[31]

Remarkable birds populated this space. The entrance was kept by feathered cherubim, who after the fall guarded its gates against human return.[32] Inside it were the "birds of paradise" whose skins were exported to sixteenth-century Europe, sometimes in the guise of parrots.[33] Those birds were often believed to be the phoenix, whose eternal life was a reward for its good behavior during Eve's tenure in the garden.[34] That bird inside the garden was, according to legend, a type of Christ, the only creature not to eat of the apple in Eve's hand.[35]

Modern retellings of the garden retained its geographic features and avian inhabitants, passing its features on to the twentieth century. Hans Christian Andersen's turn-of-the-century fairy tale "The Garden of Paradise" situates the garden and its remarkable bird on the Isle of Bliss, surrounded by a river, at the top of high cliffs beyond the Himalayas. In C. S. Lewis's *The Magician's Nephew*, the same garden was in the mountains to the west, at the crown of a hill in a green valley with a blue lake; inside the walled garden, the apple tree was guarded by a magnificent bird.[36] Hemingway set the opening of his *Garden of Eden* in the walled paradise of a small town in the Camargue, with a canal running down to the sea.[37] And in *Lilith's Brood*, by Octavia Butler, human resistors try to rebuild the failed garden high in the hills at the base of a river, in a town that they name "Phoenix" for the bird untouched by disaster.[38]

Those descriptions (minus Adam and Eve) are not unlike the typical geography of a long-dormant volcano, home of many rainforests and often, though not always, surmounted by a hidden crater lake in which floats an island, the tip of the volcano slowly rising again. Mount Bosavi, the site of Steven Feld's *Rainforest Soundwalks*, would be a strong candidate for searchers for the garden. Lush with the ancient remains of what was once a tremendous volcano, the foothills of the mountain where Feld recorded are sites characterized by running water and tropical birds. Feld's album opens with a recording of a seyak, or hooded butcherbird, singing in a miniature Eden complete with mountain and tiny Nile, "a ridge above a forest creek."[39]

A world away from this scene, a less obvious candidate for the garden lies in the Sonoran Desert, the setting of the opening page of Miyoko Chu's pop-up guide *Birdscapes*. The geography of this land is, like Mount Bosavi, the remains of volcanic formations. In this vast basin in the southwestern United States ringed by mountains, the spring rains reveal the hidden garden in the desert, when the cacti burst into bloom and a "waterfall of sound" pours from the rocky cliffs in the voice of a canyon wren.[40] Far to the north and forty-six years before Feld's recording, Olivier Messiaen, too, started his *Catalogue d'oiseaux* in an Edenic geography, the foothills of the tall peak of the Meije (or Meidje, as Messiaen calls it) near the lake of Puy-Vacher in the French Alps. In that composition, the alpine chough calls as it soars over the lake from cliff to cliff, a cold and windy parallel to the forests of Bosavi and the deserts of Sonora.

The way in which Feld, Messiaen, and Chu map species taxonomy with these idyllic topographies touches on the edges of ordered knowledge where science and artistry meet (figure 7.2). In the 1970s, while Murray Schafer explored this boundary through soundscapes, it was also being explored in the experimental

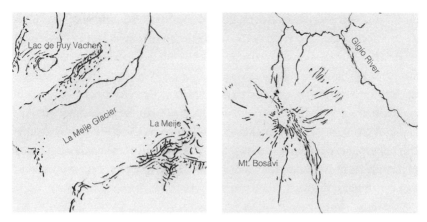

FIGURE 7.2 Mapping the structures of paradise in Feld and Messiaen:
*Left*, Lake Puy-Vacher at the foot of La Meije; *right*, crater of Mount Bosavi.

field of humanist geography, pioneered by Anne Buttimer and Yi-Fu Tuan. Tuan's attempt to merge geography with semiotics resulted in a theory of affective signs and spatial perception based on synesthesia not unlike Messiaen's compositional approach. This was a geography intended to demonstrate "how a people, not only through imaginative action but also through enlivening speech, create and recreate their multifaceted and kaleidoscopic worlds."[41] The soundscape approach to mapping sound demonstrates an affinity with this aesthetic blending of affect, anthropology, and geography.

The intentional mapping of spaces in sound calls to mind the structuralist mapping of cultural spaces. Though these maps were not really field guides, they were anthropological guides to an exotic otherness that might be seen as an appropriate foil to the guides of empirical science, for instead of locating the truth-value of formalizing structures outside of human subjectivity, they located truth-value within it. As Claude Lévi-Strauss put it in his introduction to Bororo mythology, "music has its being in me, and I listen to myself through it."[42] Lévi-Strauss's discourse on birdsong, music, and nature in his survey of Bororo myths *The Raw and the Cooked* is a forerunner of Feld's later *Sound and Sentiment*, which provides a similar summary of myths touching on birds and weeping among the Kaluli.[43] Lévi-Strauss straddles two arguments about birdsong: like Herbert Spencer, he rejects a biological tie between birdsong and human music as "hardly worth discussing"; but he also claims that birdsong is sufficiently like human music to submit to the same analysis, explaining that birdsong, like language, can be called a product of (in this case, bird) culture.[44] The analysis that follows deals with the vexing question of music's "natural" origins and attempts to outline a path in which cultural conditions, aesthetic objects, and objective knowledge might coexist. The focus of progress in Lévi-Strauss's assessment had shifted away from the human body and toward a notion of music that, in the wake of the postwar divide between science and culture, holds a residue of the science of culture that survived the reevaluations of the 1950s. In the space between birds and humans, Lévi-Strauss concluded that music was the key to this transition from disciplinary isolation to forward motion: "music is the supreme mystery of the science of man, a mystery that all the various disciplines come up against and which holds the key to their progress."[45]

## SOUNDSCAPES

Though sonic landscapes have many incarnations in twentieth-century music, Feld's *Rainforest Soundwalks*, Chu's *Birdscapes*, and Messiaen's *Catalogue d'oiseaux* are recurring themes in this chapter because of the alternative they offer to the postmodern divide against which Lévi-Strauss stalled. In their imagined landscapes, paradise is a no-place, a utopia, whose impossibility stems from the postmodern expectation that humans and birds exist in separate spheres. It is a condition that I describe in chapter 5 as the state of the postmodern human, and a condition that serves in these three cases as the point of departure for a revision of the real in which nature's purity is intact. That revision is a progressive, retro-futuristic aspiration to restore a lost balance between the human and nonhuman worlds. In each of these works, balance is restored not, as Lévi-Strauss imagined it, by the integration of disciplines, but by the integration of identity into its context. Instead of the typological pedagogy of the modern field guide, these works situate identity as a relationship experienced through a thriving natural habitat. The vitality of these (often nonhuman) places where identity lives leads back to the geography with which Messiaen, Feld, and Chu first intervened in my discussion of birdsong, the geography of the Garden of Paradise.

Messiaen's *Catalogue d'oiseaux* is, chronologically, the first of these interventions, and it comes at a time when the dominance of the spectrograph in ornithological study had only begun to be established. The *Catalogue* is one of several works marking a new period in Messiaen's oeuvre in the early 1950s, when he began to use transcriptions of birdsong as the building block of several new compositions, including *Le Merle noir*, *Reveil des oiseaux*, and *Oiseaux exotiques*. This period coincided with the growing schism between musical and scientific ways of knowing described in chapter 6, attended by the shift in Anglo ornithology toward the spectrograph and its attendant language of objectivity.

The *Catalogue*'s level of detail and the seeming exactitude of Messiaen's *style oiseaux* transcriptions have traditionally been heard through the filter provided by postwar Anglo ornithology. Though admired by scholars such as Robert Sherlaw Johnson,[46] Messiaen's birdsong transcriptions were originally met with confusion by his own students, even with laughter.[47] Declarations such as "I'm the first to have made truly scientific and, I hope, accurate notations of bird songs" and affirmations that he was an ornithologist did not assuage the reluctance of scholars to accept this church organist and harmony instructor as a professional ornithologist."[48] Messiaen was attacked by British composer Trevor Hold in the 1970s for the accuracy of his transcriptions and the "literary paraphernalia" that explained birds' habitats and species.[49] After Hold, scholars have revisited the composer's bird style many times through this lens of Anglo ornithology, agreeing on that basis that Messiaen's transcriptions are, as Peter Hill put it, "not the objective record of a scientist."[50]

Yet Messiaen situated himself in an ornithology that had little reference to that sphere, seeming to neither know nor care about the spectrographic revolution. Coming from a strong French tradition of Lamarckian evolution that emphasized habitats and ecologies over speciation, French ornithologists were not strongly invested in the Darwinian renaissance occurring at Cambridge University and in the United States. Instead, they tended to be more interested in issues of habitat, ecology, and conservation, promoting birdsong as part of a larger ecological web in need of protection from the effects of industrialized society.[51] While Thorpe's laboratory epitomized the English ornithologist of the 1950s with his graphic images and single-species research, the ornithologist of postwar France was equally a scientist and conservationist. The French counterpart to the American Ornithologists' Union (AOU) or the British BOU, for example, was a wing of the *Société d'acclimatation et de protection de la nature, La Ligue pour la protection des oiseaux* (LPO), founded in 1912.[52] The two roles of laboratory scientist and conservationist have remained part of the nation's tradition—in the 1990s, Jacques Penot suggested differentiating between the interchangeable terms *ornithologue* and *ornithologiste* to reserve "iste" for the professionally trained laboratory scientist and "ogue" for the conservationist. Penot's *ornithologue* was pointedly not an amateur, but the "most knowledgeable ones, close to the science of ornithology—the ornithologist for all the passionate people like myself [and Messiaen], who work to make birds better known and protected."[53]

When Messiaen's interest in birdsong began to intensify in the 1950s, he acted as an *ornithologue*. He attached himself to French experts who had much more in common with the fieldwork of early song collectors than with the science of the postwar British laboratory. These experts included Jacques Delamain, Robert-Daniel Etchécopar, Jacques Penot, François Hüe, and Henri Lomont.[54] Avian identity was, for these men, closely tied to lived experience in the wild. "Though the laboratory allows deeper work on certain problems" wrote Etchécopar in 1951, "the cult of managed experiences, of statistics and graphic curves, can make one lose contact with reality."[55] Jacques Delamain, probably Messiaen's most influential tutor at this time, approached birds through what Robert Fallon has called the tradition of "passionate ornithology," the nineteenth-century world of the amateur collector.[56] Delamain's birds were "les plus grandes artistes," in whom "the musical art is born from the satisfaction a being feels in translating its life into a sound."[57] While Thorpe's modern British laboratory performed focused research using short, objective recordings of isolated chaffinches, Delamain situated songbirds in relationships to one another and their habitats, relying on flowery prose to make his point:

> Under the mid-day [April] sun, the lonely phrase of an Ortolan resounded, monotone and drawling in the shadeless vineyards. With the evening breeze, sounds resumed. Then at the end of the day, the confused and artless chatter of tiny voices was extinguished. The Nightingale sang again, without conviction, in fragmented strophes. The Oriole whistled one final time. Voices after his were installed in the peace of the evening, discrete, rare, heavy with silence and darkness. A Blackbird took the stage.[58]

Messiaen's *Catalogue*, composed between 1956 and 1958 as a series of thirteen musical portraits of regional French birds and contained in seven volumes of music for solo piano, adopted this notion of the *ornithologue* as ecologist and advocate. Identity was, in these works, a reflection of habitat, context, and species relationships that mirrored Delamain's model so closely that it sometimes trespassed into plagiarism.[59] Set in gem-like geographies, each bird transcribed in the *Catalogue* was framed by details about the time of day, time of year, cohabitant species, and ecology transformed into music. These ecologies are crafted through sonic collage: in "Le Chocard des Alpes," the alpine chough calls while wheeling between the glacial rocks of the Meije peak's southwest side, set against the quick chattering of the chough (meas. 28), the polyrhythms of the glacial ascent of the mountain's side (mm. 1–26), and the gigantic ghostly arena of the

Bonne Pierre (mm. 156–87).[60] Later, in "Le Loriot," the oriole singing at dawn is bracketed by slowly ascending bass chords (mm. 1, 3, 8, 10, and so on) that culminate in the golden rays of the rising sun in parallel chords (m. 43),[61] which mimic the golden rays of the oriole's song (m. 55).[62] In Book 5, "La Bouscarle" (Cetti's warbler) moves on the borders of the Charente River one morning at the end of April, accompanied by the songs of the moorhen, kingfisher, blackbird, robin, corn crake, song thrush, wren, chaffinch, blackcap, hoopoe, nightingale, and yellow wagtail.[63]

In Messiaen's *Catalogue*, the postmodern schism between biology and aesthetics is partially mediated by the literality of rich musical ecologies. In "Le Merle de roche," for example, the score dedicated to the rock thrush unites words, images, and geography with this immediacy, opening with a "hand of rock" set in music, over which the thrush calls (figure 7.3). Messiaen's rocky hand can easily be located in images of the work's setting as a sharp, rocky fist-shaped formation for which the Cirque de Mourèze, where the work is located, is famous. Such details constructed Messiaen's own paradise, a place in which birds, separated from human disturbance, were constituted through time, through geography, through vegetation, and through their sung relationships with other birds inhabiting a shared space. Only when he turned to foreign birds did Messiaen adopt the shorter, more typological sounds of *American Bird Songs*, relying, as Robert Fallon has shown, on the sound recordings of postwar guides as his reference for the birds with whom he was less familiar.[64] But in the *Catalogue d'oiseaux*, the many species of birds, singing together in complex environments, are a sharp contrast to the short, repetitive examples of taxonomic identity that were codified in *American Bird Songs* in 1942.

From the French Alps, I turn to the representation of sonic identity in Steven Feld's *Rainforest Soundwalks*, which consciously takes on the mantle of a humanistic soundscape. Feld, who made no claims of being *ornithologiste* or *ornithologue*, has not garnered the scientific controversy to which Messiaen's work is subjected. And his recordings, unlike Messiaen's scores, have a semblance of immediate authenticity that resides in the premise of field recording. But it is no secret that these soundscapes are not unedited representations of natural sound. As EarthEar, the CD's record label, explains, "Feld shapes his academic studies of the anthropology of sound into a subtle composition that becomes an echo of the magic that's kept him coming back to his Bosavi home for a quarter century."[65] This magic is distinct from the science of nature but not from belief in the real that nature stands in for; in the words of one of Feld's reviewers, "the

FIGURE 7.3 Messiaen's score for "Le merle de roche" depicts a rock formation shaped like a fist near the Meije with a cluster chord, with "fist of rock" (*la main de pierre*) written beneath the chord, and "moonlit night—immense fist of rock, raised in a magic gesture" above it. Messiaen, *Catalogue d'oiseaux*, 1.

manipulations to which Feld subjects his sounds are subtle but unequivocal, as if 'reality' were always ready to slip into the hyper-real."[66] The hyper-reality of Feld's recordings are, in fact, the result of their unnatural technology, in which various threads of sound are edited and mixed to represent a technologically savvy rendition of the composer's sense of the border between village and forest at the foot of Mount Bosavi—neither science nor art, civilized nor wild: the garden.

Feld's "soundwalks" are intended to represent movement through just such an Edenic mixture of human and wild, though their exact trajectory is vague (and possibly imaginary, because the recordings are not literal representations of Feld walking through the forest's edge). The soundwalks combine the sounds of water, insects, and local birds—even, according to the liner notes, the sound of a fly landing on the microphone—with designations about space and time.[67] They progress from morning to afternoon and night, each track a representation of the natural diversity of the Bosavi region. "Galo," or afternoon, is a good example of this concept of a sonic "walk": it opens with various high-pitched bird calls in clear, jewel-like timbres over a background of incessantly striating insects. Several minutes into the track, one hears the cooing of a dove, a strikingly different timbre (this hoarse cooing has two distinct timbral parts, one probably a different bird, though they never overlap; the coarser and lower-pitched of these sounds continues much later into the composition). The walk ends as though traveling in a loop, with a return to the earlier cries as the dove either leaves or is left behind, and new birds cry overhead while the calls of insects grow louder again. Feld's listeners may not know the geography of the walk, but they experience an intense immersion in local sound, as if Feld has not left the habits of *Sound and Sentiment* behind. This earlier work on the Bosavi people mapped song texts onto geography extensively, and provided the reader with a general map of the area right from the beginning of the venture. And from the recording itself one knows, at least, the general landscape within which each track occurs.

Feld's naming of birds is also an indicator of his unique navigation of sonic identity, one that contrasts substantially with the use of names in Messiaen's and Chu's sonic guides. Messiaen and Chu both adopted the Latin names of natural history's taxonomies for their birds, allying their works explicitly with traditional field guides in the process. Chu's book names birds in both Latin and English, adopting unchanged the field guide tradition. Messiaen's *Catalogue* is a broader exercise in taxonomy, listing the names of species in French, Latin, English, German, and Spanish inside its front cover, and labeling birds in French within the main notation of the score. The way Messiaen names various species together in combined vocalizations and contextualization (a phenomenon that is also related to the times of day specified for each composition) suggests an attempt, however, to deploy the poetry of Delamain's guides and avoid the constrictions of the Peterson model. Unlike the exotica of *Oiseaux exotiques* and *Sept haïkaï*, the *Catalogue d'oiseaux* represents French birdsong with an environmental playfulness that filters down to the brief appearance of a diplodocus in the rocky Hérault.

Feld's naming, in contrast, adds to Latin and English taxonomies the hyperreality of the ancestral spirits that manifest in the singing bird for the Bosavi. For Feld, the idea of naming may stem in part from the influence of Murray Schafer, for whom an elaborate taxonomy of sound classifies birdsong as a type of natural sound along with the sounds of animals, sea creatures, the Apocalypse, wind, and water.[68] Yet Feld, unlike Messiaen or Chu, incorporates the element of human culture into this web of natural references. This is most evident in the opening track, "Seyak," which uses the Bosavi name for a bird Feld also calls the hooded butcherbird, *Cracticus cassicus*. Feld frames this track in the liner notes with the Bosavi belief that birds speak with the voices of the human dead. This belief, explained to Feld by his guide Yubi during the research for *Sound and Sentiment* many years before the recordings for *Soundwalks* were made, was used by Yubi's son Seyaka (whose namesake is the seyak) to suggest that the unusually drawn-out song of the seyak recorded by Feld and his assistant was actually the voice of his old guide, Seyaka's father.[69]

Feld's choice to include this anecdote brings up the unresolved taxonomic adventures of *Sound and Sentiment*, in which Feld contrasted Bosavi and Western taxonomies to highlight structural differences between the two cultures. In *Sound and Sentiment*, these differences are couched in terms of human, rather than bird, culture; in *Soundwalks*, the bird sounds as if it is left to speak for itself in a fifteen-minute soliloquy of liquid phrases. But despite the sonic impression that this "voice in the forest" is more than its Western taxonomy, Feld's written train of thought seems in conflict with his recorded one, for we are told that "*seyak* is the Bosavi name of the Hooded butcherbird, *Cracticus cassicus*," as though the bird's real name is the Latin one.[70] And in later tracks, the names of birds are given in their Western taxonomies, despite the fact that this at times conflates several sonically divergent species from across the globe into the simplified categories of scrubwrens, pigeons, and doves (while the insects are merely called insects, with the exception of the "fly" that is mentioned). Despite this, the recording of the seyak that opens the album does reconfigure a purely Western taxonomy into one that intimates a different approach, especially when the extended solo of the seyak is put in the context of the brief and momentary calls of other species in the later tracks. The effect is to open the soundwalks with a prolonged meditation on the voice of a bird.[71]

Finally, I turn to the last book in this trinity, Miyoko Chu's *Birdscapes*, which combines the ornithological aspirations of Messiaen with the recorded media that is the basis of Feld's work. This book, published in 2008, is a recent addition to a series of guides published after 2000 that incorporate tiny digital audio players into their spines, allowing users to hear birdsongs that are synced to a particular page of the book. Though they are around twice as expensive to produce as a CD audio guide, they are much easier to use and easily outsold comparable CD-based guides in the early twenty-first century. Shortly after the publication of *Birdscapes*, these guides were, at a casual measure, roughly ten times more popular on websites like Amazon.com and Barnesandnoble.com than CDs in the Peterson audio guide series, and roughly a hundred times more popular than *The Rough Guide to World Music*.[72]

The idea behind these interactive digital guides came from Becker and Mayer, a small publishing company based in Seattle that specializes in designing books with interactive features and then selling them through a larger company, in this case Chronicle Books. Becker and Mayer's first experiment integrating sound directly into a book was a publication about Marvel Comics, in which the audio component featured commentary by Stan Lee, the company's founder.[73] Becker and Mayer then focused on using the technology in audio bird guides, and worked with naturalist Les Beletsky and a team at Cornell University to create *Bird Songs*, a push-button guide to the birds of North America published in 2006.[74] This project was limited by the cost of its digital technology, and the birdsong samples were very short, about ten seconds apiece. The sound clips were picked and edited by the Cornell team to produce the most representative examples possible, the same kind of "universal" sonic identity that is the aspiration of the spectrograph examples in laboratory research and the field guide images of the Peterson system. Like so many other American guides, the source recordings for these books, including *Birdscapes*, were supplied by the audio collection at Cornell's Laboratory of Ornithology that is the source of the original recordings in *American Bird Songs*.

As data storage has gotten cheaper, Becker and Mayer has been able to increase the length of sound samples in its Birdsong series to thirty seconds, roughly the same length of comparable CD audio guides. The company's *Backyard Birdsong* series by Donald Kroodsma uses these longer examples.[75] Kroodsma is a specialist in field recording who worked much more closely with the sound samples than Beletsky did in the first series.[76] Kroodsma's sound bytes are framed by the same taxonomic project of a field guide, but he occasionally steps outside of the sonic "universals" demanded by these guides to present the listener with a changing world, as in his clip of the call-and-response pattern of male and female barred owls.[77] As digital audio files have become less expensive to produce and store, the sounds that began as guides have drifted into a decidedly more playful economy. By the early years of the twenty-first century, Americans and Europeans could purchase clocks that feature Cornell's birds chiming the hour, and a series of stuffed birds encasing tiny plastic audio files, which repeat a single two- or three-second characteristic phrase several times when squeezed.

*Birdscapes* exists close to this world of toys, for it is a book filled with elaborate pop-up habitats whose audio components are made of overlapping tracks of birds.[78] The book's jewel-toned colors and gently curved cutouts splash across gorgeous pages in an inviting fantasy of biodiversity. Though Miyoko Chu is the book's listed author, the project was a collaborative effort. Chu wrote the book's text; an illustrator, Julia Hargreaves, created illustrations that were redesigned by a small team of paper artists into three-dimensional pop-up scenes; and the book's sounds were chosen and edited by Garret Vyn, a professional wildlife recordist who drew on his own recordings as well as the Cornell audio archive at the Macaulay Library (home of the original recordings made for *American Bird Songs*).

The source recordings represent individual species, but Vyn creates an illusion of multiplicity by strategically overlapping tracks of various birds so that they seem to sing together. The tactile pleasures of the richly colored pop-up features are supplemented by speakers in the front and back covers of the book that allow a stereo playback of birds in this (somewhat limited) space. The result lasts just under thirty seconds for each of the seven "birdscapes," and the birds are both visually and sonically clustered together in space and time, roosting in a place where all hours of the day and night are compressed into less than a minute. The Sonoran Desert, for example, features a track of about twenty-seven seconds created from eleven overlapping recordings. Each bird is represented with typological repetitions of a single phrase standing in for its species. Like other soundscapes in the book, the Sonoran songs seem to be arranged with a musical effect in mind. The composition opens with a Scott's oriole's sequence of descending appoggiaturas ending in an upward turn, a quasi-baroque gesture of initiation. This opening salvo is followed by the waterfall descent of the canyon wren and several other birds, culminating two-thirds of the way through the composition with the cackling three-note descent of a Gambel's quail, set apart from the yaps of an elf owl and the low, closing trill of the nighthawk by a brief moment of silence. The result is a sonic collage that transforms the short, representative sounds of postmodern birdsong from field guide and laboratory into an object of art/science.[79]

In this setting, the only existing sights and sounds are exotic birds and spectacular vegetable landscapes (figure 7.4). This pastiche of birdsong engages the listener in a fantasy of species diversity that brings what we might call the taxonomic listening aesthetic closer to the musical idea of structural listening, where identifying many species in succession allows a kind of larger sonic identity in the form of "habitat" or, as the books suggests, "birdscape." In *Birdscapes*, nonhuman species inhabit places free of human interference, avoiding parking lots, skyscrapers, and construction sites. For the incautious viewer who fails to read the book's descriptions carefully, these birds also seem to be United States citizens, residents of the gorgeous landscapes in which they abide. Outside of their pop-up landscapes, the book's species go home every year to embattled habitats in Columbia, Brazil, Peru, Venezuela, Bolivia, and Argentina. The disenchanted reality of a global economy governed by unequal partnerships is not part of this book, however. Instead, *Birdscapes* offers a reassuring vision of ecological permanency and biodiversity that was designed, in part, as a gateway

FIGURE 7.4 The Sonoran Desert pop-up scene. Chu, *Birdscapes*.

to more difficult conversations about biodiversity, a lure to invite a comfortable public to imagine the beginnings of a better world.[80] Here is, in many ways, the ultimate expression of the postmodern human's conundrum. Nature's music is aesthetic, beautiful, and prized; but this treasure (or toy) exists in an artificial separation from the realities of the human species and postwar life. Here at last is the Eden from which Western humanity was expelled in the aftermath of racial evolutionism, a place of pleasure and ownership that does not quite exist. This final fantasy is one in which the garden is, as of old, closed to human access, but retains the birds who have not sinned.

# IN THE BEGINNING

Beyond the Oisans, the Alps of Dauphiny, over the Lac Puy-Vacher, the bird soars to the dense and greenly tangled slopes of Mount Bosavi, alighting at the border between village and forest where lies the hyper-real. But what is this garden? Some of the singing birds are stuffed decoys, others are strangely solitary, and quite a few are the products of composerly imagination. Strange though it seems, the difference between objectivity and culture is difficult to discern here. It is the imaginary birds, Feld's seyak and Messaien's *merle noir*, whose sounds acknowledge relationships with the mundane world as most human beings know it. But it is the spectral birds of bioacoustics whose calls make visible inaudible realities and reveal what is called, for lack of a better vocabulary, natural knowledge.

The natural knowledge of field guides and soundscapes bookend the problem of postmodern identity. Structured fantasies of habitat seem to offer the possibility, on the one hand, that the observer will forget herself or himself entirely and dominate difference with the Adamic power of naming the living and the dead, described in chapter 2 as the institutionalization of sonic identity. The Edens of Messiaen, Feld, and Chu are objects of white and Western privilege framed by the possibility of dystopia. Yet their fantastic landscapes also raise the possibility of escape from the unsustainable dichotomy between culture and objective reality that has constrained representations of birdsong to be only art objects, or only isolated and disembodied voices of species.

As we enter the twenty-first century, different ways of hearing identity in sound have become part of our thinking about lost Edens. For the past several decades, avian biodiversity has become a measure of environmental distress, shaping beliefs and policies about global warming and conservation.[81] The ability of amateur birders to recognize species by ear based on the pedagogy of audio field guides plays a significant role in how this measure is quantified. Annual bird counts and species lists are often a valuable source of data for larger research studies, when recorded by amateur nature lovers under the rubric of "citizen science." Most of these counts include species identified by ear, rather than by sight alone, for sound greatly increases the number of bird species that can be counted.[82] Because of this, the skill sets taught in *American Bird Song*–style audio bird guides have become a practical influence on the ways that biodiversity is defined and measured.

The practice of aural identification, however, locates the search for Eden within a problematic tradition of collecting and classifying difference in the form of songs. This tradition is part of a shared history that cannot be erased without substantial consequences. Categories of natural identity drawn from prewar typologies are the medium of legal protections for categories of difference in the twenty-first century. Our most traditional classifications of biological species, races, and cultures are those inscribed in law; they are the language that protects and promotes diversity. The shortcomings of the past have become the mechanisms of its solution. As they have always done, those who search for the Garden of Paradise must overcome difficult and complex challenges. For pilgrims of the present day, that search offers two trials that are rooted in the circumstances of the postmodern, posthuman condition: the need for an imaginative act that allows us to identify *with* aural difference beyond the scope of human uniqueness; and a historical act that allows us to identify *with* a troubled and complex past.

It is a reminder that a garden is, after all, a place that is built by labor. In Anderson's fairy tale, it was still reachable, and one could even stay there if he or she exerted sufficient self-control. Though Messiaen's, Feld's, and Chu's stories of paradise are all different, they share a musical imagining in which difference is not a kind of Occam's razor that cuts away subjectivity, and with it, those things that the voices of "others" cannot speak in linguistic terms. These geographies of birdsong are complex sonic experiences, opening pathways to Edenic futures while still shaped by the conflicts and dualities of their (and our) time. The logic of the postmodern, postwar human is part of these Edens, which sidestep important histories of inequality, racial difference, and species dualism that inform their fantasies of natural beauty and exotic difference. At the same time, these soundscapes open a space of unknown singers, a garden of imagined possibility, for each voice we do not recognize. In the midst of exoticism and human uniqueness, these are cartographies that suggest new taxonomies, raising the possibility of alternatives to the mechanisms of the past. They remind us that the past can be understood, as easily as it can be forgotten. And as maps of the garden are slowly redrawn, they hint that perhaps it will be the singing bird who finds its voice and brings us roses.

CONCLUSION

# The Animanities

I began this book with the sound of a song sparrow. A century ago, naturalist Ferdinand Schuyler Mathews heard one of these birds singing opera tunes
from Verdi's *Rigoletto*. Today, ornithologists describing the same species hear a
loud clanking sound.[1] These are striking differences: one listener hears operatic
melody, while another hears clanking. Why do these reports of the same species
differ? What does it mean to claim a shared musical identity for these two sparrows in the first place? And why would two human listeners hear that identity
in such different ways?

In this book, I have argued that the way sound is evaluated is related to the
ways animals are valued, and that sonic culture as we know it is unthinkable
without those animal lives. I began with the Darwin-Spencer debates about
animal personhood in the late nineteenth century. From that starting point, I
traced the practice of preserving and comparing songs into the natural history
museum, the field, and the laboratory. In the second half of my book, I examined
a dramatic shift in notions of sonic identity that emerged in response to postwar
reassessments of racial identity. The resulting separation between sonic studies of
human culture and animal nature continues to define aural identity in a postwar,
postmodern world. I want to conclude my book with a disruptive and unruly
intrusion upon the past century's theories of cultural identity with a made-up
discipline, "the animanities."

Una Chaudhuri once suggested that the proliferation of neologisms by animal
studies scholars is the symptom of a shared desire to radically intervene in the
established discourse of aesthetic meaning.[2] Zoontology, zoopolis, petropolis,
carno-phallogocentrism, and zooanthropology made her list of neologophilia,
in addition to zooësis, her own word at the intersection of performance studies
and animal studies.[3] Occasionally such disruptive neologisms have visited music

studies in the form of zoomusicology, ornithomusicology, and ecomusicology.[4] I add my own contribution to this "shared program of creative disciplinary disturbance" in the form of the study of living culture in a more-than-human world, the animanities.[5]

An amalgam of the *anima-* of animals and *-manities* of the humanities, the animanities are an intervention in the postwar, postmodern, posthuman condition of present-day humanism. The problems of musical knowledge described in this book are not answered by the disciplinary boundaries of the humanities. As I have suggested earlier in these pages, the dyad of science and humanism is conditioned by a history in which the subject matter of the arts and sciences has been defined against and through notions of difference as a universalizing category, a category for which the animal has been the exemplary case. In the aftermath of World War II, geneticists and anthropologists reimagined race in relation to Nazi politics, obscuring many of the ways in which difference continued to operate at the intersection of species, race, and culture. The central gesture of the animanities, then, is to reach with a laugh toward a revision of the way that notions of difference define the humanities against this limiting benchmark of human singularity.

The humanities are premised on a singularity of human identity that divides cultural knowledge from animal biology. In this book, I have located that division in postwar revisions of genetics and cultural anthropology inspired by a backlash against racism. The architects of those revisions attempted to distance intellectual work in the West from problematic racial taxonomies. But they did not address the ways that difference itself was conceptualized and measured against interconnected notions of species, race, and culture. Nor did they successfully disentangle the traditions of knowledge-making that framed personhood as contested, leaving some people as unmarked, neutral "humans" and others as deeply different.

In the past several decades, work in posthumanism has called much-needed attention to the culture/nature divide that defines the humanities. It has inspired me in my own work and in the writing of this book, and, like humanism, it is a part of my intellectual heritage. But posthumanism and its allies in object-oriented ontology, new materialism, and related trends also perpetuate the problems that I worry over in this book. As a disciplinary intervention, recent posthumanism has sometimes reinforced, rather than resisted, the erasure of racial politics from the culture/nature divide, locating itself in seemingly unmarked discursive traditions that on closer examination come from white, European, masculine voices. As a loosely defined field, posthumanism has also—despite the express wishes of many of its authors—had the effect of treating animals as if they are interchangeable with inanimate objects. In such comparisons a phonograph and a dog are treated as having comparable agency, an agency in which species and in/animacy are given importance, but race, class, gender, and other middle factors in this equation remain silent and invisible.

The thing I am calling the animanities is meant to disrupt this state of affairs. In this book, I have aspired to tell stories that give rich, specific, historical contexts to the foundations on which we build when we work within the humanities and their postscripts. By creating an imaginary field, I invite others to tell more such stories, with the belief that knowledge of our cultural and intellectual foundations is empowering. The animanities are about using specificity, richness, creativity, and context to ground the problems earmarked by posthumanism and allies in empowering histories. My goal is not to return to the question posed by vivisection, "Is knowledge worth killing for?" I do not wish to weigh the value of life against the value of knowledge; and I do not wish to weigh the value of vivisectors against the value of anti-vivisectors. Instead, I wish to create space outside the terms of those ethics of comparison by creating histories that allow us to make more informed decisions about culture and its study.

My version of this disciplinary disturbance is unabashedly aligned with the category of life, another meaning of *anima*. The life/nonlife boundary has been the subject of compelling critiques of structural inequality using the language of biopolitics and its outgrowths. When I say the animanities are aligned with life, it could seem as if I mean to dismiss these critiques, or to dismiss the ways in which "life" can be productively treated as an arbitrary and constructed category. That is not my intent. Instead, I want the animanities to foreground a set of issues that can get lost in the broad scope of nonhuman discourse: the unique demands that living beings have historically had upon ethical concerns and values. To discuss, confront, or investigate life is inextricably intertwined with ethical concerns and practices in modern society. Recent feminist and queer interventions in posthumanism have increasingly sought to interrogate the categorical limits of life as a point of entry to ethical discourse.[6] In proposing the animanities, I ask how and why we live in a sphere in which life and ethics are intertwined in the particular ways that demand this point of entry. I privilege life in order to foreground questions about how ethics circulate through living relationships, especially in the ways those relationships are defined by perceived differences. For me, inanimate objects like a kymograph do not merit the same questions as living beings like a dog, *even if the boundaries of this category of life are arbitrary.* I am less concerned with the debatable status of a virus's life than with the ethics that circulate in debates about the way the lives of nonwhite human beings, refugees, pigs, trees, or viruses are valued and evaluated. The animanities, as I imagine this project, give context to the study of human cultures, or tree cultures, or even virus cultures. But this is not, rightly or wrongly, a sphere that locates the status and rights of cups or minerals or atoms or skyscrapers on an equal basis with living beings. I do not deny the agency of inanimate objects. But I am not as concerned with the way that histories and traditions inform the rights, dignity, and personhood of the nonliving as I am with the ascription of rights, dignity, and personhood to living things.[7]

Judith Butler recently wrote that she could not imagine subjectivity without first imagining an ethical relationship as the condition of a subject's possibility.[8] Neither humanism, nor posthumanism, nor critical philosophy can fully account for the ways that such relationships have become intertwined with subjective and objective knowledge-making. I am interested in the animanities as a way to leverage histories of knowledge in order to reimagine the ways that past traditions and experiences shape what it means to think about life, what it means to assess difference, and what it means to relate ethically to others. Histories of difference and knowledge offer a foundation from which ethical choices about life are empowered and informed choices. My project of the animanities is not about providing an answer or guidebook to the ethical dilemmas of the present day; it is about providing informed contexts that enable us to confront those dilemmas more successfully.

What makes a relationship good, beneficial, right, or just? I use the word "ethical," but accounts of ethical relationships differ. For whom is a relationship good? Whom does it benefit? When is it right and just? The project I am imagining here is not about developing a single answer to these questions, but about locating the questions themselves in informed contexts that make meaningful answers possible. Which ideas, histories, traditions, and lineages give shape to an ethics of life? For whom are those ideas and traditions good, right, and just? Whose voices, interests, and hopes do they represent? For whom are these contested histories, ideas, and traditions?

In what follows, I offer a meditation on the issues that have guided my own investigation into ethical relationships in this book. Those issues emerged, for me, in the process of writing each chapter, and as a result, for each chapter of this book there is an important issue that I locate as a starting point for the animanities. Each of these issues seemed to me, in various ways, to operate along invisible fault lines at the edges of cultural life, at places where ethical decisions are formed. In the slippage at those fault lines, the disruptions around concepts such as these create a space for rethinking relationships and ethics. My hope is that by noting and problematizing those moments of instability, I can help my readers disentangle the tasks of intelligent reflection and of relating ethically to others from the taxonomies of cultural and natural difference that we inherit in the modern dyad of the sciences and humanities. If posthumanism situates its proponents through access to the nonhuman, and if animal studies is a practice of disordering the established human order, then the animanities are a more-than-human study of the histories and contexts that condition what it means to recognize and relate to those who are different.

## FAULT LINES AT THE EDGE OF LIFE

### *Personhood*

One of the first interventions of the animanities is to interrogate the values that have developed around the concept of personhood. The Darwin-Spencer debates about musical capacity that I discussed in chapter 1 offer one point of entry to the broad question of what we mean when we say someone is a person. That question is as contentious today as it was in Darwin's era. Selfhood, rights, individuality, humanity, reason, and sentience are only some of the attributes at stake in claiming that an individual is a "person." In the context of the humanities, personhood limits appropriate subject matter to the creations of human persons.

One reason the question of personhood is such a rich one is that the status of personhood is often used to determine who deserves legal rights and protections. A person is, by this definition, someone who is protected by law. In the United States and many other countries, the status of living personhood can be quite fluid, sometimes favoring corporate rather than individual persons.[9] The malleability of personhood has made it a productive terrain for legal intervention in cases of race, gender, sexuality, and other contested categories of difference. In recent years, these interventions have been the model for attempts to extend legal personhood to cetaceans and nonhuman primates, in which it is argued that certain nonhuman animals' intelligence and emotional complexity make them people who, therefore, deserve some of the legal protections of people.[10]

Yet legal personhood is a complex and two-sided coin. It confers the right to legal protection; but in so doing, it generates its opposite, the category of non-persons and illegal persons from whom rights and protection can be divested or denied.[11] The range of that denial is so broad that it begs momentary inspection: incarceration, refusal of citizenship, denial of voting rights, denial of marriage, denial of due rights or process, the relegation of nonpersons to slavery, to food production—these are among the possible fates of legal nonpersons. To have persons is to have guilt and innocence, vice and virtue.[12]

What is one to do with the radical inequality among living beings that is part of being a person? This question does not have an obvious or ready answer. The animanities invite the question, however, and suggest it is an important part of any serious rethinking of humanistic inquiry.

## *Identity*

The second intervention of the animanities is to ask how concepts of identity have institutionalized notions of difference within the humanities and the sciences. Many pages of this book are devoted to the institutionalization of identity within the context of sonic culture. Traditions of identification drawn from evolutionary theory, natural history, and mechanical quantification have all become elements in modern categories of musical style and even in the commercial branding of music's genres. The place of identity and identification in music, however, is only a small fragment of a much larger set of practices that inscribe categories of identity within the humanities and their corollaries in the sciences.

Notions of identity and identification have served many useful roles in humanistic inquiry. I am particularly interested in what historian Joan Scott called "categories of cultural analysis," categorical groups such as gender, race, or sexuality, and, by extension, cultural categories such as genre or style.[13] For Scott, such categories were "categories of analysis" precisely because they were not natural, but humanistic creations that could be subjected to historical and social analysis. Since the late 1980s, poststructural theorists, feminists, scholars of queer theory, and scholars of critical race studies have used such categories to recognize and address the flows of power that structure Western knowledge. In many cases, posthumanism has successfully extended this approach to categories of nonhuman identity.

And yet, as José Muñoz has pointed out, to think and act beyond these taxonomies is an arduous, exhausting labor.[14] Categorical analysis notes the presence of taxonomic identity, but it does not suffice to escape its institutional usage. To build relationships beyond these categories is an aspiration I share with many other scholars in animal studies, posthumanism, gender and queer studies, and anti-racist theory. There are many paths to this project, and I like to think my imaginary field the animanities will make available new trails that can be discovered, followed, tended, and enjoyed. How is life valued and evaluated with and through categories of identity, and how can it be imagined outside of history's taxonomies?

In my own attempt to escape taxonomic relationships, an important step has been locating the origins of modern categories of identity in cultural evolutionism and eugenics. This is a recognition of an erasure: I have suggested that in researchers' haste to distance intellectual labor from racial science in the period after World War II, the racial origins of categorical identity were obscured while the categories themselves remained intact. From this perspective, Scott's "categories of cultural analysis" are one step in a larger process of contextualizing modern notions of identity within histories and traditions of structural inequality.

## Difference

The institution of categorical identity is inseparable from the project of rethinking difference. Paired notions of identity and difference find a place in virtually every work of humanistic inquiry since the 1980s. They are central to research in animal studies and posthumanism; and they are constantly redefined in the sciences, as understandings of biological boundaries shift in order to reflect new discoveries and theories.[15] Through the chapters of this book, I have framed this particular pairing of identity and difference, in which institutional taxonomies of difference are the substance of categorical identity, as a unique result of the postwar, postmodern, posthuman experience.

What is difference in a postwar, postmodern, posthuman era? As an institution tied to categorical identity, it still bears a close resemblance to pre–World War II beliefs that categories such as race, gender, nation, or species had common provenance in a shared state of difference that demanded these categories remain fixed and inherent. Modern taxonomies of identity such as race or sexuality that institutionalize legal protection and rights rely on historically disparaged categories in order to protect individuals harmed by those histories. In the process, the categories themselves are protected and made necessary. It becomes strategically useful to be biologically and forever queer, biologically and forever a man or woman, biologically and forever black, or biologically and forever American, even if it is personally harmful to do so.

And personal harm and personal pleasure matter a great deal if I wish to rethink difference. In this book I have borrowed from feminist and anti-racist traditions to suggest a turn toward the kinds of evidence provided by individual experiences as an alternative approach to the taxonomies of difference. For difference is as much an experience as it is an idea; and it is generally an experience of being in an unequal relationship with someone else. If that someone is a person, then he or she is already in an unequal relationship with nonpersons; if the someone has an identity, this inequality works within institutionalized, taxonomic categories of difference.

To confront difference in this sense is not to classify difference, but to imagine it thoroughly as if it were my own experience, to identify *with* another person instead of identifying them. This act of *identification-with* can be a pleasure, perhaps the pleasure Chaudhuri means when she calls the radical otherness of animals a gift.[16] For me, this pleasure has meant (among other things) imagining myself in the place of my intellectual ancestors; imagining their inequalities and differences as part of the foundation of modern knowledge; and imagining how my own intellectual heritage is still being shaped by these intensely felt and lived unequal relationships. To treat "difference" as an act of imaginative work and play in this way is to connect the institutional dyads of humanities/science, culture/nature, and human/nonhuman to the individual experience of being deemed better or worse, more or less capable, valuable or without value, friend or food. That seemingly dire work can also be a playful and pleasurable act of forming real and potential relationships with others. The task of rethinking difference that I have set for the animanities, then, is a task of reimagining how to recognize and navigate the unequal relationships of contemporary life.

## *Knowledge*

From within those unequal relationships, I come to a question that is both radical and commonplace: Is knowledge worth killing for? This is a question with a history. In my book, I trace this history to specific traditions in laboratory physiology and comparative zoology that pitted the life of an animal against human knowledge. In those traditions, knowledge relied on the inspection and vivisection of animal bodies, an exchange of differently shaped lives for human knowledge. Animal life in this practice was an extreme of difference that made its use for food and research possible, while at the same time, animal bodies were sufficiently similar to the human body to justify research. The effect was a tradition of "intelligent reflection" in which researchers set aside their emotions in order to create knowledge through violence to animal bodies. These traditions became enormously influential in later scientific standards of objectivity and mechanical quantification in the twentieth century. In the process, however, the language and values of intelligent reflection used in hunting, vivisection, and nineteenth-century natural history were imported into the ways that objective knowledge was defined in the twentieth century.

Although machines and quantification are important to this story, its defining feature is the way that the valuation of life affected the evaluation of knowledge, as the result of a conscious decision to resist emotional responses. Quantification and mechanical objectivity are connected by this history to literal and real dilemmas about whether to use the lives of nonhuman animals in laboratory research, conflicts that are still very much a part of the laboratory biologist's work. My purpose in framing these ethical conflicts as ethical, and as evaluations of difference, reason, and emotion, is not to vilify or discredit scientists who continue to participate in this tradition. Nor is it to elevate emotion over reason, or to compare animal and human lives. Rather, my goal is to create space for conversations about ethics as *both* a needed practice with real meaning and a contextual, inherited tradition capable of being scrutinized.

The purpose of the animanities, then, is not to weigh the value of knowledge against our valuation of different lives, but to alter the terms of that comparison. The practice of the animanities is one I imagine to be informed less by the particulars of graphic notation than by the broader ethical questions that surround data formation, and the fact that those questions surround data formation. By historicizing practices of objective knowledge-making and quantification, I aspire to create a space in which the "cultural" ethics of how we make knowledge in historical, social contexts is understood to be related, and relatable, to the "scientific" ethics of truth-telling that we often call "objectivity."

## Postmodern Humanity

The animanities are a response to the postwar, postmodern, posthuman present. The conditions of being animal and being human in that moment (this moment) are the result of relatively recent divisions between culture, race, and species that date to the middle of the twentieth century. In this book, I trace those divisions to a historical, transdisciplinary backlash against racial evolutionism that occurred in the aftermath of World War II. Over a period of about twenty years, a younger generation of geneticists, anthropologists, and naturalists reimagined Darwinian evolution around an impermeable separation between genetic biology and cultural research. In biology, eugenics and typology were replaced by genetics, while anthropology was adapted to reflect a sharp division between genetic models of race and "ethnic" models of culture. One effect of this backlash against racial evolutionism, however, was that these disciplinary changes erased the visible marks of racist taxonomies, rather than confronting them. As a result, many of the tools and traditions of racial comparison persisted amid a discourse that discouraged direct discussion of "race" and, therefore, made recognition of this inheritance particularly difficult. One of the tasks of the animanities is to investigate and confront the ways in which the thing we mean by "human" in the present moment is conditioned by the erasure of race from postwar cultural and biological knowledge.

What are we to do with the sciences and the humanities, if their scope and content are shaped by a refusal to confront histories of racism? What does "racism" mean in this context, if it was (and is) entangled with the classification of species, cultures, and other gendered, classed, and historically constructed forms of difference? What does it mean to be human if the claim itself is part of this suppressed memory? What is genetics, if its terms, categories, and questions are related to prewar eugenics? What is music's study, if its styles and genres are connected to social Darwinism?

To ask these questions is not the same thing as discrediting the work of genetics or cultural research. It is not to make a claim about the personal beliefs of scientists or scholars. It does not undo or disqualify the truths uncovered by a well-crafted laboratory experiment, a thorough ethnography, or a well-researched history. It is to say that the disciplinary spheres of science and the humanities have histories of racism, speciesism, and other categorical "isms" that affect their practice. Because we have inherited relationships, communities, and connections to that past, to talk about species is to talk about race; to talk about race is to talk about culture; and to talk about culture is to talk about species. The thing I am calling the animanities is an assertion that being human in the present moment is something that depends on the interconnected histories of species, race, and culture. From this perspective, the singularity of humanism is the result of not asking, not noticing, or not recognizing the missing terms of race and other forms of difference in the division of culture from species.

Like many posthumanist authors, I find the practical work of humanism valuable and worthwhile, and I seek a critical examination of its framing devices rather than a replacement or eradication of the things that humanism does.[17] Unlike most posthumanists, however, I argue that those framing traditions are dependent on a specific problematic tradition of difference in which race has often been the middle term between species and culture. In my own work, I locate the species/race/culture trilogy within the limited time frame of the events leading up to and away from the Western backlash against eugenics and racial theory in the 1950s, and the subsequent distinction between cultural and genetic difference. Because I believe that the condition of human identity as we know it is the result of a postwar suppression of broad connections between species, race, and culture made under the rubric of categories of difference—and because those categories are still the categories under which we come to know life as different—I believe that a closer examination of the humanities demands a continued and careful examination of specific, historical notions of difference, with special attention to the way claims about animality and culture continue to be placeholders for assumptions about differences of race, gender, sexuality, and class.

## Subjectivity

How is knowledge about self and other crafted in our postwar, postmodern, posthuman world? One of the last interventions of the animanities is to examine the making of knowledge in a tradition that divides cultural studies from scientific research, and human experience from animal lives.

What is knowledge of the self and the other? When this question presented itself to postwar, postmodern European philosophers, they often turned to narratives of rupture built around the emergence of Cartesian dualism and the rise of the modern state in the seventeenth and eighteenth centuries.[18] Those narratives, though compelling, were distanced from the twentieth-century taxonomies of race and species that had contributed to human identity in the postwar era. More recently, a few scholars have pointed toward untold stories of racism and speciesism that seem to explain aspects of the methodological gap between humanistic and scientific epistemology.[19] In my own work, I have found the intertwined histories of race and species to be particularly useful points of entry to understanding modern notions of subjectivity and objectivity. To know one's self and its others, for me, has meant getting to know the category we call "animal."

In my book, I explored the problem of self and other by examining a division between music scholars and ornithologists over the meaning and value of objective knowledge in the 1970s. In studies of song, the differing values applied to animal and human lives led to striking differences in the way sonic knowledge and evidence were conceived, created, and evaluated. These differences were of particular importance in the sphere of the laboratory, where today's standards of evidence are tied to traditions of objectivity rooted in the nineteenth century's desire for emotional distance under the guise of an "intelligent reflection" that distinguished between animal and human life. By inventing the animanities, I want to raise questions about how we are supposed to know things about ourselves and others when we live in the shadow of this tradition, in which the tools for knowledge have developed amid oppositional ways of valuing emotion, constructing evidence, and evaluating life in the sciences and the humanities.

## Paradise

I conclude my seven themes with the idea that it is worth imagining what it is like to be someone very different from who you (think you) are, in a place very different from where you (think you) live. The history presented in this book is paralleled by a century of commerce and urbanization that has yielded lost landscapes, altered songs, and the exoticization of nature in comfortably far-off places. In the midst of those changes, it has become unclear if there is a natural "other," or if nature is a place one can go to in any meaningful way. The theory of the anthropocene, the geologic age of human impact, recognizes the extremity of these lost landscapes, but it does so by embracing the false neutrality of the postwar, postmodern, postracial human animal, in the name of the "anthropos."[20] In such a world, one could argue that Western intellectual practitioners have much to learn about imagining alternate selfhood.

Those whose songs have been primarily heard, and sung, through taxonomies of difference—in this book alone, a group that includes all singing birds, women, deaf zoologists, Angolans, and many others—are vast in number. In chapter 7 I turned to three attempts to imagine alternate selfhood through the songs of birds, set within fantastical natural landscapes conditioned by the postmodern human animal. Each of those works came from deep within music's taxonomies, building a Garden of Eden whose characters were exotic and exoticized as well as beautiful. They pose both a hope and a problem. A hope, that in imagining the songs of those who are radically different from ourselves, we can have a hand (and ear) in building a garden in which joy, beauty, and dignity replace the violence that often defines the unequal relationships of difference. A problem, in that those unequal relationships are reinforced when the garden is imagined through the exclusionary inheritance of a postwar, posthuman, postracial human identity.

In locating my last intervention in the spectacularly unusual setting of paradise, I return to the thought that diverse ethics and imagination are at the core of the project I am calling the animanities. We live in a century in which nations, peoples, places, species, and we ourselves have been violently exploited or exploitative, harmful or harmed, subjected or objectified. The labor of extricating ourselves from this world asks that we reexamine the ethics, values, and historical lineage that got us here. The soundscapes of Messiaen, Feld, and Chu represent a product of postmodern rupture, but they are not the only Edens of the postwar period. Intellectuals ranging from Donna Haraway to Aph and Syl Ko have pointed to counternarratives of paradise, particularly those crafted by Afro-futurist artists, that resist the seeming neutrality of the postwar human animal.[21] Such visions of paradise allow us to confront the ways in which even our hopes are defined by categorical inequalities, and to wonder in what world one might be truly satisfied to trade one's current lot with an animal or some other "other."

Of old, the humanities were the secular counterpart to religious scholasticism in Europe's tradition, an arena in which individuals learned the knowledge and skills that made them ethical citizens. The work of imagining paradise in creative, thorough, and material ways positions the animanities to support and strengthen ethical traditions of global citizenship founded on new ways of thinking about the values of life and difference. Like the medievals, in my book I try to imagine paradise as a literal and earthly place that, just possibly, could exist.

## ENDINGS AND BEGINNINGS

I mean for this book to be the beginning of something—an intervention in the ethics of humanism that I hope is followed by changing conversations, by questions and ideas, and by departures I myself cannot foresee. I hope the animanities are a *something* that can be a useful tool in music studies; useful to historians of science, animal studies scholars, and posthumanists. I would be delighted by praise, gratitude, and success. But the animanities are not a paradigm shift, a great invention, or a foundational moment. To make this so would perpetuate the conditions under which difference leaks into the present from the nineteenth century to define success as well as knowledge in bodily terms.

There are many reasons to change those terms. The postmodern human animal, divided between culture and biology, has begun to take new material forms defined by its schismatic world. The largely forgotten history of American eugenics, separated after World War II from the history of biology, has been linked by recent historians of science to the emergence of elements of genetic research within the American for-profit healthcare industry, and to the ongoing attempts that have existed in international science since World War II to classify race genetically.[22] In 2013, scientists genetically modified the DNA of human embryos, initiating the first stage of an international contest to genetically engineer human beings, with the question of race still framed by a daunting ignorance about the history of scientific measures and ideals of difference.[23] A world away from those laboratories, the genome is also the frame of music in the digital marketplace, framed as a genetic entity whose body is the basis of Pandora radio's classification of music genres, the "music genome," a sonic typology that monetizes the methodology of the song collectors of the early twentieth century.[24]

The challenge of modern knowledge, like the ethics of Hornbostel's laboratory, need not be a comparison of life and difference, or a weighing of the values of the humanities against the values of science. Yet the postmodern human animal does not inhabit this world comfortably. Is genetic research in the twenty-first century served by a world in which scientific objectivity and cultural knowledge are divided? Is the economy of music served by that science, inherited from the nineteenth century? Can the Edens of our future be created in this divided world? I do not think so. Nor do I think we can reach for those Edens without confronting the burdens of our unequal relationships, a heritage whose ethical navigation I present as both hard work and adventurous play.

The structure of the animanities that I have mapped in this conclusion is meant to serve as a preliminary path beyond those dichotomies. Its structures, drawn from my own proclivities, will be better informed by the voices of a community than the work of any single individual. I offer this book as a creative beginning, a start, and a question to which others must offer a more complete answer.

# NOTES

## Introduction

1. Ferdinand Schuyler Mathews, *Field Book of Wild Birds and Their Music* (New York: Putnam, [1904] 1921), 116.

2. See Ulric Daubeny, "Gramophone 'Why Nots?,'" *Musical Times* 61, no. 929 (July 1920): 486; Lawrence Gilman, "Music of the Month: A Tone Poet from Italy," *North American Review* 215, no. 795 (Feb. 1922): 267.

3. Agnes Huplitz, "Poliomyelitis and Cranberries," *American Journal of Nursing* 17, no. 6 (Mar. 1917): 504.

4. Henry Oldys, "Music of Man and Bird," *Harper's Monthly* 114 (Apr. 1907): 771.

5. Albert R. Brand, "A Method for the Intensive Study of Bird Song," *Auk* 52, no. 1 (Jan. 1935): 40–52.

6. Aretas Saunders, "The Song of the Song Sparrow," *Wilson Bulletin* 63, no. 2 (June 1951): 101.

7. The associations of "Reuben" with minstrelsy were still remembered in the twentieth century, where it was featured, among other things, as an interlude in performances of *Uncle Tom's Cabin*. Edward Morrow, "Poor Old Uncle Tom," *Prairie Schooner* 4, no. 3 (Summer 1930): 178; 174–80.

8. "Song Sparrow," Wikipedia, accessed July 25, 2016, https://en.wikipedia.org/wiki/Song_sparrow#Song.

9. Excellent examples of such canon-making descriptions of sound in animal behaviorism include D. Graham Burnett, *The Sounding of the Whale: Science and Cetaceans in the Twentieth Century* (Chicago: University of Chicago Press, 2012); Joeri Bruyninckx, "Sound Sterile: Making Scientific Field Recordings in Ornithology," in *The Oxford Handbook of Sound Studies*, ed. Karin Bijsterveld and Trevor Pinch (New York: Oxford University Press, 2011), 127–50; Gregory Radick, *The Simian Tongue* (Chicago: University of Chicago Press, 2007).

10. Jonathan Sterne, *The Audible Past: Cultural Origins of Sound Reproduction* (Durham, NC: Duke University Press, 2003); Mara Mills, "Deaf Jam: From Inscription to Reproduc-

tion to Information," *Social Text* 28, no. 1 (Spring 2010): 35–58; Julia Kursell, "A Gray Box: The Phonograph in Laboratory Experiments and Fieldwork, 1900–1920," in Bijsterveld and Pinch, *Oxford Handbook of Sound Studies*, 176-197.

11. The story of Edison's ear phonautograph is told in Sterne, *The Audible Past*, 31.

12. See, for example, Alexandra Hui, *The Psychophysical Ear: Musical Experiments, Experimental Sounds, 1840–1910* (Cambridge, MA: MIT Press, 2013); Benjamin Steege, *Helmholtz and the Modern Listener* (New York: Cambridge University Press, 2012).

13. Una Chaudhuri and Holly Hughes, "Introduction," in *Animal Acts: Performing Species Today* (Ann Arbor: University of Michigan Press, 2014), 2.

14. See, for example, Colin Dayan, *With Dogs at the Edge of Life* (New York: Columbia University Press, 2016); Zakiyyah Iman Jackson, "Outer Worlds: The Persistence of Race in Movement 'Beyond the Human,'" *GLQ* 21, no. 2–3 (June 2015): 215–18; Clare Jean Kim, *Dangerous Crossings: Race, Species, and Nature in a Multicultural Age* (New York: Cambridge University Press, 2015); Kalpana Rahita Seshadri, *HumAnimal: Race, Law, Language* (Minneapolis: University of Minnesota Press, 2012); Alexander Weheliye, *Habeus Viscus: Racializing Assemblages, Biopolitics, and Black Feminist Theories of the Human* (Durham, NC: Duke University Press, 2014).

15. Bruno Latour, *We Have Never Been Modern*, trans. Catherine Porter (Cambridge, MA: Harvard University Press, 1993); Donna Haraway, *Primate Visions: Gender, Race, and Nature in the World of Modern Science* (New York: Routledge, 1989); Katherine Hayles, *How We Became Posthuman: Virtual Bodies in Cybernetics, Literature, and Informatics* (Chicago: University of Chicago Press, 1991).

16. George Santayana, *The Life of Reason or The Phases of Human Progress: Reason in Common Sense* (New York: Scribner, [1905] 1920), 284.

17. Ibid., 284–85.

18. See note 14.


**ONE**  *Why Do Birds Sing? And Other Tales*

1. Henry W. Oldys, "A Remarkable Hermit Thrush Song," *Auk* 30 (Oct. 1913): 541. Oldys (born Henry Worthington Olds) didn't argue that birdsong was the direct predecessor of human music in the evolutionary timeline, but rather that they mapped parallel aesthetic developments.

2. Robert Lach, *Studien zur Entwickelungsgeschichte der ornamentalen Melopöie* (Lepizig: C. F. Kahnt Nachfolger, 1913), 545: "dieser Frage, die für die musikalische Entwickelungsgeschichte darum von größter Bedeutung ist, weil sie den Schlüssel zum Problem der Entstehung von Sprache und Musik bildet."

3. Peter Bowler, *Evolution: The History of an Idea* (Berkeley: University of California Press, 1989), 285.

4. See Julian Huxley, *Evolution: The Modern Synthesis*, 2nd ed. (London: Allen and Unwin, 1963), 22–24. See also *The Evolutionary Synthesis: Perspectives on the Unification of Biology*, ed. Ernst Mayr and William B. Provine (Cambridge, MA: Harvard University Press, 1980); Bowler, *Evolution*; or Vassilliki Betty Smocovitis's more recent "'It Ain't Over 'Til It's Over': Rethinking the Darwinian Revolution," *Journal of the History of Biology* 38, no. 1 (Spring 2005): 33–49.

5. An excellent summary of their broader differences as evolutionists can be found in Mark Francis, *Herbert Spencer and the Invention of Modern Life* (New York: Routledge, 2007). Many thanks to Gregory Radick for suggesting this source.

6. In this chapter and in chapter 5, I reproduce some of the details first published in Rachel Mundy, "Evolutionary Categories and Musical Style from Adler to America," *Journal of the American Musicological Society* 67, no. 3 (Fall 2014): 735–68.

7. See August Weismann, "The Musical Sense in Animals and Men," *Popular Science Monthly* 37 (July 1890): 352–58. See also "Music for Animals," *Musical Times and Singing Class Circular* 38, no. 649 (Mar. 1, 1897): 162–63; and even much later, Esther Goetz Gilliland, "The Healing Power of Music," *Music Educators Journal* 31, no. 1 (Sept.–Oct. 1944): 18–20.

8. Ivan Pavlov, in "New Researches on Conditioned Reflexes," *Science* 58, no. 1506 (Nov. 9, 1923): 360; and in "The Scientific Investigation of the Psychical Faculties or Processes in the Higher Animals," *Science* 24, no. 620 (Nov. 16, 1906): 616.

9. George Herzog, "Do Animals Have Music?," *Bulletin of the American Musicological Society* 5 (Aug. 1941): 3–4; Edward Ellsworth Hipsher, "Do Animals Like Music?," *Musical Quarterly* 12, no. 2 (Apr. 1926): 166–74.

10. Herzog, "Do Animals Have Music?," 4.

11. In addition to the examples cited here, I recommend David Rothenberg's *Why Birds Sing: A Journey into the Mystery of Bird Song* (New York: Basic Books, 2006), which contains a wonderful survey of early examinations of animal music.

12. Rev. Samuel Lockwood, "A Singing Hesperomys," *American Naturalist* 5, no. 12 (Dec. 1871): 764.

13. Lockwood, "Singing Hesperomys," 761–70; Charles Darwin, *The Descent of Man, and Selection in Relation to Sex* (New York: Barnes and Noble, [1871] 2004), 514–16; Joseph Sidebotham, "Singing Mice," *Nature* 17, no. 418 (Nov. 8, 1877): 29; George J. Romanes, "Singing Mice," *Nature* 17, no. 418 (Nov. 8, 1877): 29.

14. There were six editions of Cornish's book between 1894 and 1914, with a positive review in the science journal *Nature* in 1895. Charles John Cornish, *Life at the Zoo: Notes and Traditions of the Regent's Park Gardens* (London: Seeley, 1896).

15. John Glen described older Scottish tunes as "the spontaneous product of natural melody, irrespective of any established principles whatever," in Glen's *Early Scottish Melodies* (Edinburgh: J. and R. Glen, 1900), 3.

16. Cornish, *Life at the Zoo*, 122–27.

17. James Sully, "Animal Music," *Cornhill* 40 (Nov. 1879): 609.

18. Grant Allen, "Aesthetic Evolution in Man," *Mind* 5, no. 20 (Oct. 1880): 445–64; Richard Wallaschek, *Primitive Music* (New York: Longman, Green, 1893); Francis H. Allen, "The Aesthetic Sense in Birds as Illustrated by the Crow," *Auk* 36, no. 1 (Jan. 1919): 112–13; Aretas A. Saunders, "Review: The Song of the Wood Pewee," *Auk* 61, no. 4 (Oct. 1944): 658–60.

19. Francis H. Allen, "The Evolution of Bird-Song," *Auk* 36, no. 4 (Oct. 1919): 531.

20. Charles Darwin, *The Expression of the Emotions in Man and Animals*, (London: John Murray, 1872), 87.

21. "Review: The Power of Sound, by Edmund Gurney," *British Quarterly Review* 73, no. 145 (Jan. 1881): 209.

22. Athanasius Kircher, *Musurgia universalis sive ars magna consoni et dissoni, in X libros digesta*, vol. 1 (Rome: Francesco Corbelletti, 1650), 27, 30; see also Suzannah Clark and Alexander Rehding, "Introduction," in *Music Theory and Natural Order from the Renaissance to the Early Twentieth Century*, ed. Suzannah Clark and Alexander Rehding (New York: Cambridge University Press, 2001), 1–14.

23. Herbert Spencer, "The Origin and Function of Music" in *Essays: Scientific, Political, and Speculative*, vol. 2 (New York: Appleton, 1907), 426. See also Kerman and Tomlinson's description of the role of music in nineteenth-century culture in Joseph Kerman and Gary Tomlinson, *Listen*, 7th ed. (Boston: Bedford/St. Martin's, 2012), 209.

24. For discussions of music's spiritual powers among Herder and his contemporaries, see Mary Sue Morrow, *German Music Criticism in the Late Eighteenth Century: Aesthetic Issues and Instrumental Music* (New York: Cambridge University Press, 1997); see also, for example, Andrei Bely, *Gogol's Artistry*, trans. Christopher Colbath (Evanston, IL: Northwestern University Press, 2009).

25. Arthur Conan Doyle, "A Study in Scarlet," in *Sherlock Holmes: The Complete Novels and Stories*, vol. 1, introduction by Loren Estleman (New York: Bantam Books, [1887] 1986), 33.

26. W. L. Goodwin, "The Nature of Music and the Music of Nature," *Queen's Quarterly* 11, no. 2 (Oct. 1903): 197.

27. "Status Approuvés par Décision Minisérielle du 8 Mars 1866," *Bulletin de la Société de Linguistique de Paris* 1 (1871): 12.

28. Henri Bergson, *Creative Evolution*, trans. Arthur Mitchell (New York: Holt, 1913), 158.

29. Michel Lejeune, who wrote the entry on language and writing in *L'évolution humaine des origines à nos jours*, vol. 3, ed. M. Lahy-Hollebeque (Paris: Librairie Aristide Quillet, 1934), wrote "L'étape par laquelle on passe de l'animal, qui ne parle pas, à l'homme, animal parlant, est donc en définitive l'accession à une intellectualité supérieure" (296),

making the same distinction as Bergson in defining animal signs as fixed (and therefore, in Lejeune's analysis, not properly language).

30. See Gregory Radick, *The Simian Tongue* (Chicago: University of Chicago Press, 2007).

31. "Music for Animals," 163.

32. Chauncey J. Hawkins, "Sexual Selection and Bird Song," *Auk* 35, no. 4 (Oct. 1918): 421.

33. Darwin, *Descent of Man*, 514–16.

34. Ibid., 519.

35. Sully, "Animal Music," 605–21.

36. Charles A. Witchell, *The Evolution of Bird-Song, with Observations on the Influence of Heredity and Imitation* (London: Adam and Charles Black, 1896), 60; see also 12–21.

37. Alfred Russel Wallace, *Darwinism: An Exposition of the Theory of Natural Selection* (New York: Macmillan, 1889), 284; see also Alfred Russel Wallace and James Marchant, *Letters and Reminisces* (New York: Harper, 1916), 246–48.

38. Weismann, "Musical Sense in Animals and Men," 352.

39. "Review: The Power of Sound," 209.

40. Spencer, "Origin and Function of Music," 362.

41. Darwin, *Expression of the Emotions*, 90.

42. Ibid.

43. Ibid., 87.

44. Jules Combarieu, *La Musique, Ses Lois, son Évolution* (Paris: Flammarion, 1907), 161; Sully, "Animal Music," 607; Edmund Gurney, *The Power of Sound* (London: Smith, Elder, 1880), 476; see also Edmund Gurney, "On Some Disputed Points in Music," *Fortnightly Review* 20, no. 115 (July 1876): 106–30.

45. Gurney, *Power of Sound*, 479.

46. Herbert Spencer, "The Origin of Music," *Mind* 15, no. 60 (Oct. 1890): 29, 450.

47. Ibid., 458.

48. Ibid., 29, 452–54.

49. As naturalist Francis Allen asserted about birdsong, animal music signified "*social*, instead of biological, evolution." Francis H. Allen, "Group Variation and Bird-Song," *Auk* 40, no. 4 (Oct. 1923): 646.

50. Sully, "Animal Music," 619.

51. Grant Allen, "Aesthetic Evolution in Man," 445.

52. See Xenos Clark, "Animal Music, Its Nature and Origin," *American Naturalist* 13, no. 4 (Apr. 1879): 201; Witchell, *Evolution of Bird-Song*.

53. Bowler, *Evolution*, 9; Timothy Ingold, *Evolution and Social Life* (New York: Cambridge University Press, 1986), 26.

54. See Mark Pluciennik, *Social Evolution* (London: Duckworth, 2005), 32–40. Plucien-

nik also offers a compelling comparison of what he calls the racialized "stadial schemes" of nineteenth-century social evolutionism with the Enlightenment "universalism" that preceded them in the eighteenth century.

55. Weismann, "Musical Sense in Animals and Men," 358. Weismann was not referring to a biological racial difference, but to a difference in cultivation or "soul," as he put it.

56. Henry Oldys, "Music of Man and Bird," *Harper's Monthly* 114 (Apr. 1907): 771.

57. Cornish, *Life at the Zoo*, 125.

58. This "old-fashioned" approach was most famously advocated by Thomas Carlyle. His essays on the heroic can be found in *On Heroes, Hero-Worship, and the Heroic in History*. Music histories that relied on Carlyle's approach include Houston Stewart Chamberlain's *Richard Wagner* and many of his other writings on music, or the biographical approach to music history in general texts such as Theodore Baker's *Biographical* and John Knowles Paine et al.'s multivolume American series *Famous Composers and Their Works*. I discuss this at greater length in my article "Evolutionary Categories and Musical Style from Adler to America," *Journal of the American Musicological Society* 67, no. 3 (Fall 2014): 735–68.

59. Herbert Spencer, *The Study of Sociology* (New York: Appleton, [1878] 1901), 30. I first encountered the phrase "great man history" in feminist criticism, but Spencer seems to have used it much earlier in his attacks on Thomas Carlyle in *The Study of Sociology*.

60. Spencer, "Origin and Function of Music," 424.

61. Spencer, "Origin of Music," 425–26.

62. See, for example, Wallaschek, *Primitive Music*; Charles Hubert Parry, *The Evolution of the Art of Music*, ed. H. C. Colles (New York: Appleton, 1893); W. J. Treutler, "Music in Relation to Man and Animals," *Proceedings of the Musical Association*, 25th Session (1899): 71–91; Combarieu, *La Musique, Ses Lois, son Évolution*; Carl Stumpf, *Die Anfange der Musik* (Lepizig: Barth, 1911).

63. Guido Adler, "Internationalism in Music," trans. Theodore Baker, *Musical Quarterly* 11, no. 2 (Apr. 1925): 282. Adler's theory actually offered a subtle reply to some racist typologies, for he argued that musical hybrids, like hybridized livestock and crops, were superior to isolated musical styles. He claimed that his own city of Vienna epitomized these superior musical crossbreeds.

64. Guido Adler, "'The Scope, Method, and Aim of Musicology,' (1885): An English Translation with an Historico-Analytical Commentary," trans. and commentary Erica Mugglestone, *Yearbook for Traditional Music* 13, no. 19 (1981): 7.

65. Guido Adler, *Der Stil in der Musik* (Leipzig: Breitkopf and Härtel, 1911), 142.

66. Parry, *The Evolution of the Art of Music*, 61.

67. Ibid., 4–5.

68. Lach, *Studien zur Entwickelungsgeschichte der ornamentalen Melopöie*, 539–42.

69. Ibid., 543–44.

70. Daniel Gregory Mason, *The Art of Music*, vol. 1 (New York: National Society of Music, 1915), 8.

71. Witchell, *Evolution of Bird-Song*, 48, 58.

72. Simeon Pease Cheney, *Wood Notes Wild: Notations of Bird Music* (Boston: Lee and Shepard, 1891), 5.

73. Francis Allen, "The Evolution of Bird-Song," 530.

74. Sully, "Animal Music," 621.

75. Grant Allen, "Aesthetic Evolution in Man," 447.

76. Henry Home, Lord Kames, *Elements of Criticism*, 9th ed., ed. Abraham Mills (New York: F. J. Huntington and Mason, [1762] 1855), 15.

77. Marjorie Garson, *Moral Taste: Aesthetics, Subjectivity and Social Power* (Buffalo, NY: University of Toronto Press, 2007), 6.

78. See Charles Harrison, Paul Wood, Jason Gaiger, eds., *Art in Theory, 1815–1900: An Anthology of Changing Ideas* (Malden, MA: Blackwell, 1998); Edward Lippman, *A History of Western Musical Aesthetics* (Lincoln: University of Nebraska Press, 1992).

79. John Finney, *Music Education in England, 1950–2010: The Child-Centered Progressive Tradition* (Burlington, VT: Ashgate, 2011), 33.

80. Spencer, "Origin and Function of Music," 425.

81. Ibid., 424.

82. Wallaschek, *Primitive Music*, 243.

83. Richard Wallaschek and James McKeen Cattell, "On the Origin of Music," *Mind* 16, no. 63 (July 1891): 379, 381.

84. Stumpf, *Die Anfang der Musik*.

85. Conwy Lloyd Morgan, *An Introduction to Comparative Psychology* (New York: Scribner, [1894] 1896), 59.

86. "The animal 'word,' if we like so to term it, is an isolated brick; a dozen, or even a couple of hundred such bricks do not constitute a building . . . It is just because language is the expression of a portion of a scheme of thought that it indicates in the speaker the possession of a rational soul." Conwy Lloyd Morgan, *Animal Behaviour* (London: Edward Arnold, 1900), 205.

87. T. J. McCormack, "Review: *An Introduction to Comparative Psychology*," *Monist* 5, no. 3 (Apr. 1895): 443; "Review: *An Introduction to Comparative Psychology*," *British Medical Journal* 1, no. 1788 (Apr. 1895): 762.

88. Radick, *The Simian Tongue*, 52. Radick's book offers a rich and contextualized history of behaviorism that situates it in relation to animal vocalization research.

89. Morgan, *Introduction to Comparative Psychology*, 366.

90. Ibid., 362.

91. Ibid., 366.

92. Sully, "Animal Music," 605.

93. Oldys, "Music of Man and Bird," 766.

94. William Gardiner, *The Music of Nature* (Boston: Wilkins and Carter, 1841), 19; Cheney, *Wood Notes Wild*, 3; Gene Stratton Porter, *Music of the Wild* (Jennings and Graham, 1910), 59.

95. The examples of this are numerous and well documented. Two starting points are Bruno Nettl, *The Study of Ethnomusicology: Thirty-One Issues and Concepts* (Champaign: University of Illinois Press, 2005); and Gregory F. Barz and Timothy J. Cooley, eds., *Shadows in the Field: New Perspectives for Fieldwork in Ethnomusicology*, 2nd. ed. (New York: Oxford University Press, 2008).

96. In addition to his work on birds, Mathews also did work with the Gray Herbarium at Harvard. See T. S. Palmer, "Obituaries: Ferdinand Schuyler Mathews," *Auk* 57, no. 1 (Jan. 1940): 138.

97. Ferdinand Schuyler Mathews, *Field Book of Wild Birds and Their Music* (New York: Putnam, 1921), xxxix.

98. Ibid., 254–62.

99. Oldys, "The Rythmical Song of the Wood Pewee," 272. "Swanee River" is also attributed to Edwin Christy, but Oldys understood this as a Foster song.

100. Palmer, "Obituaries," 616–17.

101. See, for example, Wallace Craig, "The Twilight Song of the Wood Pewee: A Preliminary Statement," *Auk* 43, no. 2 (Apr. 1926): 150–52; Francis Allen, "Review: The Song of the Wood Pewee"; Aretas A. Saunders, "Review: The Song of the Wood Pewee," *Auk* 61, no. 4 (Oct. 1944): 658–60.

102. Saunders, "Review: The Song of the Wood Pewee," 659.

103. Sully, "Animal Music," 605.

104. Weismann, "The Musical Sense in Animals and Men," 357, 358.

105. Treutler, "Music in Relation to Man and Animals," 83.

106. Wallaschek, "On the Origin of Music," 381.

107. William Wallace, *The Threshold of Music* (New York: Macmillan, 1908), 6.

108. Robert Moore, "The Fox Sparrow as a Songster," *Auk* 30, no. 2 (Apr. 1913): 180, 182.

109. Francis Allen, "Aesthetic Sense in Birds as Illustrated by the Crow," 113.

110. Richard Hunt, "Evidence of Musical 'Taste' in the Brown Towhee," *Condor* 24, no. 6 (Nov.–Dec. 1922): 198.

111. Fiona Erskine, "The Origin of Species and the Science of Female Inferiority," in *Charles Darwin's The Origin of Species: New Interdisciplinary Essays*, ed. David Amigoni and Jeff Wallace (New York: Manchester University Press, 1995), 95–121.

112. See Elizabeth Grosz, *The Nick of Time: Politics, Evolution, and the Untimely* (Durham, NC: Duke University Press, 2004); and Kimberly Hamlin, *From Eve to Evolution: Darwin, Science, and Women's Rights in Gilded Age America* (Chicago: University of Chicago Press, 2014).

113. Hawkins, "Sexual Selection and Bird Song," 421–37. The Nineteenth Amendment, granting women the right to vote, began to circulate the House and Senate in 1918 and was eventually ratified in 1920.

114. Ibid., 428.

115. Ibid.

116. Saunders, "Evolution of Bird Song," 149–51; Francis Allen, "The Evolution of Bird-Song," 528–36.

117. Francis Allen, "Group Variation and Bird-Song," 644.

118. Saunders, "Evolution of Bird Song," 150.

119. Ibid. See also Nathan Pieplow, "Subsong vs. Whisper Song," *Earbirding*, http://earbirding.com/blog/archives/2910 (accessed Oct. 16, 2011), for a more recent account and comments on whisper singing.

**TWO** *Collecting Silence: The Sonic Specimen*

1. See *Library of Congress Strategic Plan 2011–2016* (Washington, DC: Library of Congress, Nov. 2010), 4.

2. I don't know yet how it will be classified, because I am still writing it.

3. Jann Pasler, "The Utility of Musical Instruments in the Racial and Colonial Agendas of Late Nineteenth-Century France," *Journal of the Royal Musical Association* 129, no. 1 (2004): 24–76. Organologist F. W. Galpin, who worked with collections at both the American Museum of Natural History and the Metropolitan Museum of Art, seems to have considered the instruments at the AMNH to be organized according to the racial categories created by J. W. Powell at the Bureau of American Ethnology (F. W. Galpin, "The Whistles and Reed Instruments of the American Indian of the North-West Coast," *Proceedings of the Musical Association* 29 [1902]: 115). Franz Boas, who organized many of those original instruments in the museum, adopted contemporary ethnological classifications. The Crosby-Brown collection at the Metropolitan Museum of Art, with which the current collection originated in 1889, organized non-Western instruments by their nation or region of origin, and organized Western instruments (with which the original curators were more familiar) by their presumed evolutionary relationships. Mary Elizabeth Brown, *Preliminary Catalogue of the Crosby-Brown Collection* (New York: Metropolitan Museum of Art, 1901), 5–14.

4. Harlan I. Smith, "Museums of Sounds," *Science* 40, no. 1025 (Aug. 1914): 274.

5. For a thorough discussion of specimens and cultural evolutionism in anthropology, see Timothy Ingold, *Evolution and Social Life* (New York: Cambridge University Press, 1986).

6. See "difference" and "identity" in James Champlin Fernald, *The Concise Standard Dictionary of the English Language* (New York: Funk and Wagnalls, 1910), 122, 211; Edward

T. Roe, ed., *Laird and Lee's Webster's New Standard American Dictionary* (Chicago: Laird and Lee, 1911), 594; Harry Thurston, ed., *New Websterian 1912 Dictionary* (New York: Syndicate, 1912), 253, 431; *Webster's Collegiate Dictionary*, 3rd ed. (Springfield, MA: Merriam, 1916), 490; George J. Hagar, ed., *The New Supreme Webster Dictionary* (New York: Adair and Petty, 1919), 381.

7. H. W. Fowler and F. G. Fowler, eds., *The Concise Oxford Dictionary of Current English* (New York: Oxford University Press, 1911), 229, 403.

8. Thomas Henry Huxley, *Introduction to the Study of Zoology: The Crayfish*, 4th ed. (London: Kegan Paul, Trench 1880), 291. See also George Sandeman, *Problems of Biology* (London: Swan Sonnenschein, 1896), 15, 30.

9. Sandeman, *Problems of Biology*, 31.

10. To cite just a handful: Guy West Wilson, "The Identity of Mucor Mucedo," *Bulletin of the Torrey Botanical Club* 33, no. 11 (Nov. 1906): 557–60; Ralph Cürtiss Benedict, "The Type and Identity of Dryopteris Clintoniana (D. C. Eaton) Dowell," *Torreya* 9, no. 7 (July 1909): 133–40; M. L. Fernald, "The Identity of Angelica Lucida," *Rhodora* 21, no. 248 (Aug. 1919): 144–47; C. A. Robbins, "The Identity of Cladonia Beaumontii," *Rhodora* 29, no. 343 (July 1927): 133–38.

11. Lorraine Daston, "Type Specimens and Scientific Memory," *Critical Inquiry* 31, no. 1 (Autumn 2004): 174.

12. G. F. Ferris, *The Principles of Systematic Entomology*, Stanford University Biological Sciences, series 5, no. 3 (San Francisco: Stanford University Press, 1928), 153.

13. George William Hunter, *Elements of Biology* (New York: American Book, 1907).

14. Donna Haraway, "Teddy Bear Patriarchy: Taxidermy in the Garden of Eden, New York City," *Social Text* 11 (Winter 1984–85): 20–64; Mieke Bal, "Telling, Showing, Showing Off," *Critical Inquiry* 18, no. 3 (Spring 1992): 556–94. Haraway describes the museum's inscriptions as having the "unmistakable themes" of "youth, paternal solicitude, virile defense of democracy, and intense connection to nature." Haraway's analysis offers a rich history of individual specimens' origins and their transformation from living animals to exhibited skins (21).

15. National Museum of Natural History, "A History of the Department of Anthropology, 1897–1997," accessed June 18, 2015, http://anthropology.si.edu/outreach/depthist.html.

16. Library of Congress, "About the American Folklife Center," accessed June 18, 2015, http://www.loc.gov/folklife/aboutafc.html.

17. Jesse Walter Fewkes, "On the Use of the Phonograph in the Study of the Languages of American Indians," *Science* 15, no. 378 (May 2, 1890): 268. I first read this quote in Erika Brady, *A Spiral Way: How the Phonograph Changed Ethnography* (Jackson: University of Mississippi Press, 1999), 55, a valuable survey of the phonograph in ethnography that contains many examples of what I call sonic "specimen collecting."

18. Mary P. Winsor, *Reading the Shape of Nature: Comparative Zoology at the Agassiz*

*Museum* (Chicago: University of Chicago Press, 1991), 202. Winsor's book has an excellent summary of the dates during which many of the Agassiz students were at the Newport laboratory, and their often stormy relations with Agassiz (200–211).

19. Fewkes was accused of plagiarizing, and had used data for one of his articles from work he had compiled at Newport; Agassiz, as the lab's director, would have expected to have the authorial rights to this work. See Winsor, *Reading the Shape of Nature*, 208–9; see also Steve Ayers, "Jesse Walter Fewkes: to Preserve and Protect," *Camp Verde Bugle*, Mar. 13, 2008, accessed June 17, 2015, http://campverdebugleonline.com

20. The Cornell recordings culminated with the first commercial recorded guide to birdsong, A. A. Allen and P. P. Kellogg's *American Bird Songs* (Ithaca, NY: Comstock, 1942); and Cowell's stash is described in Michael Hicks, *Henry Cowell, Bohemian* (Champaign: University of Illinois Press, 2002), 119.

21. Aaron Glass, "Frozen Poses: Hamat'sa Dioramas, Recursive Representation, and the Making of a Kwakwaka'wakw Icon," in *Photograph, Anthropology, and History: Expanding the Frame*, ed. Elizabeth Edwards and Christopher Morton (Burlington, VT: Ashgate, 2009), 94.

22. Benjamin Ives Gilman, "The Science of Exotic Music," *Science* 30, no. 772 (Oct. 15, 1909): 532.

23. Brady, *A Spiral Way*.

24. See Charles K. Wead, "The Study of Primitive Music," *American Anthropologist* 2, no. 1 (Jan. 1900): 77; Franz Boas, "Teton Sioux Music," *Journal of American Folklore* 38, no. 148 (1925): 319.

25. Percy Grainger, "The Impress of Personality in Unwritten Music," *Musical Quarterly* 1, no. 3 (July 1915): 423.

26. The Metropolitan Museum of Art's collection of musical instruments was begun with a group of instruments donated by Mary Crosby-Brown in 1889; the Musiksinstrumenten-Museum Berlin was similarly founded in 1888; the Stearns collection was donated by Frederick Stearns to the University of Michigan in 1898; the Victoria and Albert Museum acquired instruments through the mid-1800s, forming the bulk of its collection around the 1882 purchase of a large private collection assembled by Carl Engel; and the Paris Conservatoire considerably expanded its collection in the 1870s under Gustave Chouquet (and, much later, André Schaeffner reconstituted another large collection in Paris in 1928 as an ethnological collection at the Musée de l'Homme).

27. The Metropolitan Museum of Art's collection is filled with animal-shaped instruments; I recall the curator, Ken Moore, telling me in 2001 that these animal instruments had curb appeal for early collectors and continue to charm visitors today.

28. I learned this while updating the original 1889 Crosby-Brown catalogue as an intern there in 2001.

29. Curt Sachs, "La signification, la tâche et la technique muséographique des collections d'instruments de musique," reprinted from *Mouseion* 27/28 in *Cahiers de musiques traditionelles* 16 (2003): 33.

30. Curt Sachs and Erich M. von Hornbostel, "Classification of Musical Instruments," trans. Anthony Baines and Klaus P. Wachsmann, *Galpin Society Journal* 14 (Mar. 1961): 8–9.

31. Pasler, "Utility of Musical Instruments."

32. See Michael V. Pisani, "'I'm an Indian Too': Creating Native American Identities in Nineteenth- and Early Twentieth-Century Music," in *The Exotic in Western Music*, ed. Johnathan Bellman (Dexter, MI: Northwestern University Press, 1998), 218–57; Simeon Pease Cheney, *Wood Notes Wild: Notations of Bird Music* (Boston: Lee and Shepard, 1891), 3.

33. In addition to the examples I provide here, a valuable discussion of this fascination with scientific notation in the field of ornithology is provided by Joeri Bruyninckx, "Sound Science: Recording and Listening in the Biology of Bird Song, 1880–1980" (PhD diss., University of Maastricht, 2012).

34. Ferdinand Schuyler Mathews, *Field Book of Wild Birds and their Music* (New York: Putnam, [1904] 1921), viii, 308–12; American Museum of Natural History, Clarence Weed Correspondence MS W444, Folder 1904–1910, Ferdinand Schuyler Mathews to Clarence Weed, Jan. 28, 1904.

35. Mathews, *Field Book of Wild Birds*, viii, 308–12.

36. Lucy V. Baxter Coffin, "Individuality in Bird Song," *Wilson Bulletin* (June 1928): 97–99.

37. See, for example, Benjamin Ives Gilman, *Hopi Songs* (Cambridge, MA: Riverside Press, 1908); Aretas Saunders, "Some Suggestions for Better Methods of Recording and Studying Bird Songs," *Auk* 32, no. 2 (Apr. 1915): 173–83; Frances Densmore, *Teton Sioux Music* (Washington, DC: Smithsonian Institution Bureau of American Ethnology Bulletin 61, 1918). The topic's relevance to ornithology has been discussed by Joeri Bruyninckx in "Sound Science," and by myself in Rachel Mundy, "Birdsong and the Image of Evolution," *Society and Animals* 17, no. 3 (2009): 206–23.

38. Albert Brand, "A Method for the Intensive Study of Birdsong," *Auk* 52 (Jan. 1934): Plate 6. Bruyninckx's "Sound Science" documents a striking shift away from amateur fieldwork and toward standardization that was promoted by Brand and others like him. It's a story that fascinates me as well, for these changes retrospectively labeled one practice "amateur" and the other "science."

39. Otto Abraham and Erich von Hornbostel, "Vorschläge zur Transkription exotischer Melodien," *SIMG* 2, no. 1 (1909–10): 1–25.

40. Frances Densmore, *Chippewa Music II* (Washington, DC: Smithsonian Institution Bureau of American Ethnology Bulletin 53, 1913).

41. Gilman, *Hopi Songs*, 21.

42. Several examples of this are given in Valérie Chansigaud's *Histoire de l'ornithologie* (Paris: Delachaux et Niestlé, 2007), 110–18.

43. This process of comparison was described and debated in numerous texts; see, for example, Julius MacLeod's *The Quantitative Method in Biology* (New York: Longman, Green, 1919), which prefigures the eventual emergence of the evolutionary synthesis in its critique of disciplinary disparities in comparative study.

44. One introduction to biology explained this concept to readers with an analogy that was both insightful and revealing of its time: "Without any idea of classifying in a scientific way, we are constantly noting the characteristics of people we see; and we compare and distinguish between people of different races (white, black, red, etc.), of nations (German, English, French, Swedish, etc.), of people from different regions of a country (Northern, Southern, and Western in the United States), and of individuals in families." Anna Nieglieh Bigelow and Maurice Alpheus Bigelow, *Applied Biology: An Elementary Textbook and Laboratory Guide* (New York: Macmillan, 1917), 139.

45. See Paul J. Korshin, *Typologies in England, 1650–1820* (Princeton, NJ: Princeton University Press, 1982).

46. Peter Harrison, *The Bible, Protestantism, and the Rise of Natural Science* (New York: Cambridge University Press, 1998), 129.

47. Edward Eigen, "Overcoming First Impressions: Georges Cuvier's Types," *Journal of the History of Biology* 30, no. 2 (Summer 1997): 179–209.

48. Georges Cuvier, *The Animal Kingdom, Arranged According to Its Organization*, trans. H. McMurtrie (London: Orr and Smith, 1834), 192, 426.

49. Eigen, "Overcoming First Impressions," 190.

50. Professor Coues and S. S. Oregon, "On Some New Terms Recommended for Use in Zoological Nomenclature," *Auk* 1, no. 4 (Oct. 1884): 321.

51. Charles Schuchert and S. S. Buckman, "The Nomenclature of Types in Natural History," *Science* 21, no. 545 (Jun. 9, 1905): 900.

52. As Thone's discourse on modern taxidermy's capacity to match skins to their ideal form explains, "True beauty, whether in a woman or a lioness, goes clear to the bones." Frank Thone, "Not 'Stuffed,'" *Science News-Letter* 28, no. 752 (Sept. 7, 1935): 154.

53. Haraway, "Teddy Bear Patriarchy."

54. Phillips Barry, "Folk-Music in America," *Journal of American Folklore* 22, no. 83 (1909): 78; Lucy Broadwood and Cecil J. Sharp, "Some Characteristics of English Folk-Music," *Folklore* 19, no. 2 (June 1908): 132–34.

55. Aretas Saunders, "The Song of the Field Sparrow," *Auk* 39, no. 3 (July 1922): 386–99; Aretas Saunders, "The Song of the Song Sparrow," *Wilson Bulletin* 63, no. 2 (June 1951): 99–109.

56. Albert Rich Brand, "The 1935 Cornell-American Museum Ornithological Expedition," *Scientific Monthly* 41, no. 2 (Aug. 1935): 187–90.

57. Laura Boulton Collection, Archives for Traditional Music, University of Indiana. Correspondence I 32 B 21 (Laura Boulton to Albert Brand, Dec. 5, 1932); I 35 B 124 (Laura Boulton to Albert Brand, Nov. 29, 1935).

58. See, for example, Gilman, *Hopi Songs*; Densmore, *Teton Sioux Music*; Helen Roberts, "New Phases in the Study of Primitive Music," *American Anthropologist* 24, no. 2 (Apr.–Jun. 1922): 144–60.

59. Densmore, *Teton Sioux Music*, 53; Roberts, "New Phases in the Study of Primitive Music," 151.

60. Charles Hubert Parry, *Style in Musical Art* (London: Macmillan, 1911), 163, 166, 322; see also Guido Adler, "Internationalism in Music," trans. Theodore Baker, *Musical Quarterly* 18 (1932): 289.

61. Cecil James Sharp and Charles L. Marson, *Folk Songs from Somerset* (London: Simpkin, 1905), xii.

62. Carl E. Seashore, "Psychology in Music: The Role of Experimental Psychology in the Science and Art of Music," *Musical Quarterly* 16, no. 2 (Apr. 1930): 234.

63. To mention only a few examples of this vast literature, see Wead, "Study of Primitive Music"; Percy Grainger, "Collecting with the Phonograph," *Journal of the Folk-Song Society* 3, no. 12 (1908): 148; Gilman, "Science of Exotic Music"; Ray E. Miller, "A Strobophotographic Analysis of a Tlingit Indian's Speech," *International Journal of American Linguistics* 6, no. 1 (Mar. 1930): 49; William J. Entwistle, "Notation for Ballad Melodies," *PMLA* 55, no. 1 (Mar. 1940): 61–72.

64. Eugenie Lineff, *The Peasant Songs of Great Russia*, second series (Moscow: Municipal Printing Office, 1912), 61.

65. Roberts, "New Phases in the Study of Primitive Music," 148.

66. Bruno J. Strasser, "Collecting Nature: Practices, Styles, and Narratives," *Osiris* 27, no. 1 (2012): 329.

67. I compiled this comparison by paraphrasing the more extended comparisons Gene Stratton-Porter made in *Moths of the Limberlost* (New York: Doubleday, 1916), 15–20; and *Music of the Wild* (Cincinnati: Jennings and Graham, 1910), 55.

68. Stratton-Porter, *Music of the Wild*, 59.

69. Peter Bartók, *My Father* (Tampa, FL: Bartók Records, 2002), 213.

70. Laszlo Somfai's study of this movement mentions that Bartók's family heard crickets and frogs in the work, an association that he believed was suggested by Bartók himself. Laszlo Somfai, "Analytical Notes on Bartók's Piano Year of 1926," *Studia Musicologica Academiae Scientiarum Hungaricae* 26, Fasc. 1/4 (1984): 6.

71. *Miért és hogyan gyüjtsünk népzenét?* [Why and How Do We Collect Folk Music?]

(Budapest: Somló Béla, 1936), trans. in *Béla Bartók, Essays*, ed. Benjamin Suchoff (Lincoln, NE: Bison, 1992), 19–20.

72. See, for example, Julie Brown, *Bartók and the Grotesque: Studies in Modernity, the Body, and Contradiction in Music* (Burlington, VT: Ashgate, 2007). This otherwise excellent reference on Bartók and bodies does not mention a single beetle, butterfly, mosquito, cricket, or other kind of insect.

**THREE** *Collecting Songs: Avian and African*

1. Suzanne Cusick, *Francesca Caccini at the Medici Court: Music and the Circulation of Power* (Chicago: University of Chicago Press, 2009), xvi.

2. As in the écriture féminine championed by Hélène Cixous in "The Laugh of the Medusa," trans. Keith Cohen and Paula Cohen, *Signs* 1, no. 4 (Summer 1976): 875–93, and the influence it has had on books such as Judith Tick's landmark music biography *Ruth Crawford Seeger: A Composer's Search for American Music* (New York: Oxford University Press, 1997).

3. Ruth Behar, "Introduction," in Ruth Behar and Deborah A. Gordon, *Women Writing Culture* (Berkeley: University of California Press, 1995), 4.

4. See Helen Myers, ed., *Ethnomusicology: Historical and Regional Studies* (New York: Norton, 1993); and Joeri Bruyninckx, "Sound Science: Recording and Listening in the Biology of Bird Song, 1880–1980" (PhD diss., University of Maastricht, 2012).

5. See Bálint Sarosi, "Hungary and Romania," in *Ethnomusicology*, ed. Myers, 187–96; see also Eugenie Lineff, *The Peasant Songs of Great Russia*, second series (Moscow: Municipal Printing Office), 1912; Cecil J. Sharp, "Folk-Song Collecting," *Musical Times* 48, no. 767 (Jan. 1907): 16–18. In Japan, the musician-scholar Jin Nyodo was one of several intellectuals who collected and recorded Japanese traditional music in the early decades of the twentieth century; he is known for collecting and transcribing *honkyoku*, or Zen flute music, from different regions in Japan's mainland. For animal vocalization collections, see Alwin Voigt, *Exkursionsbuch zum Studium der Vogelstimmen: Praktisch Anleitung zum Bestimmen der Vögel nach irhem Gesange* (Leipzig: Quelle and Meyer, 1909); F. de Fenis, *Les Langages des Animaux et de l'homme* (Hanoi: Extrême-orient), 1921; Walter Garstang, *Songs of the Birds* (London: John Lane, 1922).

6. A useful overview of this tradition is provided in Erika Brady, *A Spiral Way: How the Phonograph Changed Ethnography* (Jackson: University of Mississippi Press, 1999). See also Cecil James Sharp and Charles L. Marson, *Folk Songs from Somerset* (London: Simpkin, 1905), xiii; Sharp, "Folk-Song Collecting"; Deborah Scarborough, *A Song Catcher in Southern Mountains: American Folk Songs of British Ancestry* (New York: Columbia University Press, 1937); Laura Boulton, *The Music Hunter* (New York:

Doubleday, 1969); Anonymous, "British Folk-Song," *Musical Times* 45, no. 733 (Mar. 1904): 175–76; Anonymous, "The Collecting of English Folk-Songs," *Musical Times* 46, no. 746 (Apr. 1905): 258.

7. Wallace Craig, "Request for Data on the Twilight Song of the Wood Pewee," *Science* 63, no. 1638 (May 1926): 525.

8. He also put a second advertisement in *Auk*. Wallace Craig, "The Twilight Song of the Wood Pewee: A Preliminary Statement," *Auk* 43, no. 2 (Apr. 1926): 150–52.

9. See Correspondence Regarding the Wood Pewee, New York State Archives, Box A 3157–55.

10. The two most thorough biographical memoirs available on Craig are the material in Richard Burkhardt's *Patterns of Behavior: Konrad Lorenz, Niko Tinbergen, and the Founding of Ethology* (Chicago: University of Chicago Press, 2005), 34–61; and Theodora J. Kalikow and John A. Mills, "Wallace Craig (1876–1954), Ethologist and Animal Psychologist," *Journal of Comparative Psychology* 103, no. 3 (1989): 281–88.

11. Wallace Craig, "The Voices of Pigeons Regarded as a Means of Social Control," *American Journal of Sociology* 14, no. 1 (July 1908): 99.

12. An introduction to the field of disability studies can be found in Lennard J. Davis, ed., *The Disability Studies Reader*, 4th ed. (New York: Routledge, 2013). This chapter is not primarily about disability studies, but Wallace's situation clearly points toward a broader need to deconstruct power and notions of the "normal" body. Yet he was also an exceptional case in certain ways, losing his hearing late in life, with very little money for medical support or reeducation; and he seems to have identified more strongly with eugenicists than with the deaf community.

13. Alexander Agassiz, *Letters and Recollections of Alexander Agassiz*, ed. George Russell Agassiz (New York: Houghton Mifflin, 1913), 390.

14. Mary P. Winsor, *Reading the Shape of Nature: Comparative Zoology at the Agassiz Museum* (Chicago: University of Chicago Press, 1991), 200–202.

15. Charles B. Davenport, "The Personality, Heredity, and Work of Charles Otis Whitman, 1843–1910," *American Naturalist* 51, no. 601 (Jan. 1917): 5–30.

16. May 2, 1901, Wallace Craig to Charles Adams, Western Michigan University, University Archives and Regional History Collections, Charles C. Adams: Folder 2, Wallace Craig Correspondence.

17. Charles C. Adams, "Introductory Note: Bird Song and the State Museum," in Wallace Craig, *The Song of the Wood Pewee* Myochanes virens linnaeus: *A Study of Bird Music* (*New York State Museum Bulletin* 334, June 1943), 7. See also May 11, 1901, Wallace Craig to Charles Adams. Western Michigan University, University Archives and Regional History Collections, Charles C. Adams: Folder 2, Wallace Craig Correspondence.

18. "Turtur risorius?" asks the reader deeply invested in dove taxonomy. The bird does not exist in modern ornithology, and it seems that Whitman may have made his

own version of the nomenclature, which today uses the name *Streptopelia risoria* to describe the ringneck doves that would have included *Turtur risorius*—though that nomenclature, too, is contested. See Charles Otis Whitman, *Inheritance, Fertility, and the Dominance of Sex and Color in Hybrids of Wild Species of Pigeons: Posthumous Works of Charles Otis Whitman*, ed. Oscar Riddle (Washington, DC: Carnegie Institute, 1919), 189.

19. Mark V. Barrow, *Nature's Ghosts: Confronting Extinction from the Age of Jefferson to the Age of Ecology* (Chicago: University of Chicago Press, 2009), 125.

20. Wallace Craig, "A Note on Darwin's Work on the Expression of the Emotions in Man and Animals," *Journal of Psychology and Social Psychology* 16, no. 5 (Apr. 1921): 362.

21. Craig, "Voices of Pigeons," 100. See also Wallace Craig, "The Expression of Emotion in the Pigeons. II. The Mourning Dove (Zenaidura macruora Linn.)," *Auk* 28, no. 4 (Oct. 1911): 401.

22. Craig, "Voices of Pigeons," 100.

23. Wallace Craig, "The Expression of Emotion in the Pigeons. III. The Passenger Pigeon (Ectopistes migratorius Linn.)," *Auk* 28, no. 4 (Oct. 1911): 408–27.

24. Craig, "Expression of Emotion in the Pigeons. III," 408.

25. James O. Dunn, "The Passenger Pigeon in the Upper Mississippi Valley," *Auk* 12, no. 4 (Dec. 1895): 389.

26. Burkhardt, *Patterns of Behavior*, 47.

27. Wallace Craig, "The Expression of Emotion in the Pigeons. I. The Blond Ring-Dove," *Journal of Comparative Neurology and Psychology* 19, no. 1 (Apr. 1909): 32.

28. Craig, "Expression of Emotion in the Pigeons. III," 408.

29. *Maine, A Guide "Down East"* (State of Maine WPA, Federal Writers' Project, 1937), 132.

30. May 24, 1913, Wallace Craig to Charles Adams, Western Michigan University, University Archives and Regional History Collections, Charles C. Adams: Folder 2, Wallace Craig Correspondence.

31. Craig to Yerkes July 30, 1923, Robert Mearns Yerkes Papers Box 12, Series I: Professional Correspondence, Folder 208, Sterling Memorial Library Manuscripts and Archives, Yale University.

32. Margaret W. Rossiter, "Philanthropy, Structure, and Personality or, The Interplay of Outside Money and Inside Influence" in *Science at Harvard University: Historical Perspectives*, ed. Clark A. Elliott and Margaret W. Rossiter (Cranbury, NJ: Associated University Presses, 1992), 20.

33. Richard Farnum, "Patterns of Upper-Class Education in Four American Cities: 1875–1975," in *The High Status Track: Studies of Elite Schools and Stratification*, ed. Paul William Kingston and Lionel S. Lewis (Albany: State University of New York, 1990), 57.

34. Farnum, "Patterns of Upper-Class Education in Four American Cities," 58.

35. Rossiter, "Philanthropy, Structure, and Personality," 21.

36. Susan Burch, *Signs of Resistance: American Deaf Cultural History, 1900 to World War II* (New York: New York University Press, 2002), 156–57.

37. Quoted in Winsor, *Reading the Shape of Nature*, 199.

38. Wallace Craig to Robert Yerkes, May 15, 1923, Robert Mearns Yerkes Papers Box 12, Series I: Professional Correspondence, Folder 208, Sterling Memorial Library Manuscripts and Archives, Yale University. For a brief summary of Yerkes's ties to eugenics, see Daniel J. Kevles, *Eugenics and the Uses of Human Heredity* (Berkeley: University of California Press, 1985), 80–82.

39. Wallace Craig to Charles B. Davenport, Aug. 8 1920, Charles Davenport Papers, Mss. B. D27, Series I, American Philosophical Society Library.

40. Charles Benedict Davenport, *The Science of Human Improvement by Better Breeding* (New York: Holt, 1910), 32.

41. Wallace Craig to Leonard Carmichael, Feb. 3, 1936, Craig Correspondence, Leonard Carmichael Papers, American Philosophical Society, Mss B. C212.

42. Anne E. MacLoghlin to Wallace Craig, Sept. 10 [?], 1926, Folder 2, Correspondence Regarding the Wood Pewee, New York State Archives, Box A 3157–55.

43. Mrs. Delbert G. Lean to Wallace Craig, Aug. 27, 1929, Folder 1, Correspondence Regarding the Wood Pewee, New York State Archives, Box A 3157–55.

44. Craig, "The Twilight Song of the Wood Pewee: A Preliminary Statement," 119–22.

45. Ibid.

46. Ernst Mayr to Charles Adams, Sept. 24, 1943. Western Michigan University, University Archives and Regional History Collections, Charles C. Adams: Folder 2, Wallace Craig Correspondence.

47. Leonard Carmichael to Miss Noonan, Aug. 6, 1945, Craig Correspondence, Leonard Carmichael Papers, American Philosophical Society, Mss B. C212.

48. Wallace Craig to Robert Yerkes, Mar. 1, 1944, Robert Mearns Yerkes Papers Box 12, Series I: Professional Correspondence, Folder 208, Sterling Memorial Library Manuscripts and Archives, Yale University.

49. Carl E. Seashore to Charles Davenport, May 12, 1919; Charles Davenport to Carl E. Seashore, Jan. 5, 1920, Box 85, Carl E. Seashore folder, Charles Davenport Papers, Mss. B. D27, Series I, American Philosophical Society Library.

50. Charles Davenport to Carl E. Seashore, Apr. 15, 1919, Box 85, Carl E. Seashore folder, Charles Davenport Papers, Mss. B. D27, Series I, American Philosophical Society Library. See also Nathaniel Comfort, *The Science of Human Perfection: How Genes Became the Heart of American Medicine* (New Haven: Yale University Press, 2012), 40.

51. Charles Davenport, "Comparative Social Traits of Various Races," *School and Society* 14, no. 340 (1921): 344–48 (reprinted in 1923 with Laura T. Craytor listed as coauthor, "Comparative Social Traits of Various Races," *Journal of Applied Society* 7 [1923]: 127–34).

The children were grouped as Italian, Irish, German, English, Swedish, Russian, Austrian, Jewish, Greek, and African.

52. Charles Davenport to Laura Craytor, Sept. 12 1922, Box 110, Laura Craytor Boulton Folder, Charles Davenport Papers, Mss. B. D27–2, Series II Subseries B. Records of Assistants, American Philosophical Society Library.

53. Charles Davenport to Frederick Lucas, Sept. 12, 1922, Correspondence Box I 22 B, Laura Boulton Collection, Archives for Traditional Music, University of Indiana. See also Correspondence Box I 24 B.

54. See chapter 3, "Botanizers, Birders, and Other Naturalists," in Glenda Riley, *Women and Nature: Saving the "Wild" West* (Lincoln: University of Nebraska Press, 1999), 43–61.

55. Bruno J. Strasser, "Collecting Nature: Practices, Styles, and Narratives," *Osiris* 27, no. 1 (2012): 329.

56. Laura Boulton Collection, Angola Box 2, Folder 3.

57. Boulton, *Music Hunter*, 58.

58. Alzada Carlisle Kistner, *An Affair with Africa: Expeditions and Adventures across a Continent* (Washington, DC: Island, 1998), 124.

59. Linda Marinda Heywood, *Contested Power in Angola, 1840s to the Present* (Rochester, NY: University of Rochester Press, 2000), 14–18.

60. Mike Stead, Sean Rorison, and Oscar Scafidi, *Angola* (Guildford, CT: Bradt Travel Guides, 2013), 4–16.

61. Boulton, *Music Hunter*, 59.

62. Ibid., 61.

63. A classic text on this topic is Donna Haraway's *Primate Visions: Gender, Race, and Nature in the World of Modern Science* (New York: Routledge, 1989).

64. See American Museum of Natural History Central Archives, Box 1213, Folder 1928, Nov.–Dec. Straus correspondence.

65. Boulton, *Music Hunter*, 62.

66. Carnegie Museum of Natural History, 1931 Pulitzer Angola Expedition 2006-8, Box 1.

67. Boulton, *Music Hunter*, 66–67.

68. Ibid., 3.

69. Ibid.

70. See Laura Boulton Collection, Archives for Traditional Music, Africa-Angola Box 2 of 3, Folder 1.

71. "I still wear the wide copper one [bracelet] she gave me with its geometrical designs," she writes in *The Music Hunter* (72), and in later photographs she does appear to be wearing a wide copper bracelet with geometric designs.

72. Actually, Rudyerd initially named it *Seicercus laurae*; in line with ornithology's constantly shifting nomenclatures, the genus was later changed to *Phylloscopus*.

73. Laura Boulton Collection, Archives for Traditional Music, Africa-Angola Box 2 of 3, Folder 1.

74. Laura Boulton Collection, Archives for Traditional Music, Series IIIC Research, Folder 3; Lecture Presentation Notes–Africa.

75. Laura Boulton Collection, Archives for Traditional Music, Africa-Angola, IIIA Box 2 Folder 1, Ethnographic notes on Ovimbundu.

76. Laura Boulton Collection, Archives for Traditional Music, Africa-Angola Box 2 of 3, Folder 1, 3.

77. Harry A. Allard, *Our Insect Instrumentalists and Their Musical Technique* (Washington, DC: United States Government Printing Office, from the Smithsonian Report for 1928, 1929), 591.

78. Ethnomusicologists have criticized this economy since the mid-1990s, and today many of Laura's recordings are in the process of being repatriated to their original communities with the collaboration of ethnomusicologists Aaron Fox and Chie Sakakibara.

**FOUR** *Songs on the Dissecting Table*

1. Erich M. von Hornbostel, "Formanalysen an siamesischen Orchesterstücken," *Archiv für Musikwissenschaft* 2, no. 2 (Apr. 1920): 320. I first encountered this quote in Eric Ames, "The Sound of Evolution," *Modernism/modernity* 10, no. 2 (Apr. 2003): 314, and I mention it in my article "Evolutionary Categories and Musical Style from Adler to America," *Journal of the American Musicological Society* 67, no. 3 (Fall 2014): 737–70. I owe much to Ames's essay, which I read years ago, forgot, and rediscovered as I worked on this chapter. I am also extremely grateful for the generous support of Karin Bijsterveld and the University of Maastricht, where I spent time as a visiting researcher in 2012 studying the history of transcription from 1900 to 1940 with Bijsterveld and Joeri Bruyninckx.

2. Jaap Kunst, "Zum Tode Erich von Hornbostel's," *Anthropos* 32, no. 1/2 (Jan.–Apr. 1937): 239.

3. The university changed names several times in the nineteenth and twentieth centuries. Originally the University of Berlin, it was renamed Friedrich-Wilhelms University in 1828, and Humboldt University in 1945. Sources often refer to it colloquially as "The University of Berlin," and I have adopted this in order to make it clear that I'm referring to a single institution. However, its proper name during Hornbostel's tenure was Friedrich-Wilhelms University.

4. This was a dilemma that brings the laboratory epistemology made familiar by Bruno Latour and Steve Woolgar in *Laboratory Life* (Los Angeles: Sage, 1979) into conversation with the ethical challenges of weighing the value of actual lives. It is not objects who are the central "actors" in this chapter, but the lives of animals and other "others" that they come to represent.

5. *Prohibiting Vivisection of Dogs: Hearings Before the Subcommittee of the Committee on the Judiciary United States Senate* (Washington, DC: Government Printing Office, United States, 1919), 84.

6. William Norman Guthrie, "A New Star," *Sewanee Review* 12, no. 3 (July 1904): 356; Paul Bechert, "The Donaueschingen Festival," *Musical Times* 66, no. 991 (Sept. 1925): 845.

7. Wallace Craig, "The Expression of Emotion in the Pigeons. I. The Blond Ring-Dove," *Journal of Comparative Neurology and Psychology* 19, no. 1 (Apr. 1909): 32.

8. Karl H. Eschman, "The Rhetoric of Modern Music," *Musical Quarterly* 7, no. 2 (Apr. 1921): 163.

9. O. G. Sonneck, "Music for Adults and Music for Children," *Music Supervisors' Journal* 15, no. 2 (Dec. 1928): 23.

10. William Arms Fisher, "What is Music?," *Musical Quarterly* 15, no. 3 (July 1929): 362–63.

11. Basil Maine, "Shaw, Wells, Binyon—And Music," *Musical Quarterly* 18, no. 3 (July 1932): 377–78.

12. Edward J. Dent, "The Romantic Spirit in Music," *Proceedings of the Musical Association*, 59th Sess. (1932–33): 95.

13. Philip E. Vernon, "The Apprehension and Cognition of Music," *Proceedings of the Musical Association*, 59th Sess. (1932–33): 62.

14. See Mary Ann Elston, "Women and Anti-Vivisection in Victorian England, 1870–1900," in *Vivisection in Historical Perspective*, ed. Nicolaas A. Rupke (New York: Croom Helm, 1987), 259–95.

15. *Prohibiting Vivisection of Dogs*, 82.

16. Ibid., 83.

17. "The Vivisection Question in Germany," *North Carolina Medical Journal* 4, no. 1, (July 1879): 40.

18. Rudolf Virchow, "Über den Werth des pathologischen Experiments," *Transactions of the International Medical Congress*, vol. 1 (London: Kolckmann, 1881), 34.

19. See Ulrich Tröhler and Andreas-Holger Maehle, "Anti-Vivisection in Nineteenth-Century Germany and Switzerland: Motives and Methods," in *Vivisection in Historical Perspective*, ed. Nicolaas A. Rupke (New York: Croom Helm, 1987), 149–87.

20. Carl Stumpf, "Das Psychologische Institut," in G. Lenz (Hrsg.), *Geschichte der königlichen Friedrich-Wilhelm Universität*, vol. 3 (Halle: Verlag der Buchhandlung des Waisenhauses, 1910), 202–7.

21. Mitchell G. Ash, "Academic Politics in the History of Science: Experimental Psychology in Germany, 1879–1941," *Central European History* 13, no. 3 (Sept. 1980): 272–73. See also Mitchell G. Ash, *Gestalt Psychology in German Culture 1890–1967: Holism and the Quest for Objectivity* (New York: Cambridge University Press, 1998).

22. Stumpf expressed concern about being seen as another Wundt, "I am in any case

of the opinion that large-scale research in experimental psychology has objective difficulties as well . . . for my part I could not decide, now or later, to follow the example of Wundt and the Americans in this direction." Quoted in Ash, "Academic Politics in the History of Science," 273.

23. An American student, Herbert Langfield, summarized Stumpf's seminar in 1937 using notes from his participation in the 1906–1907 winter semester. See Herbert S. Langfield, "Stumpf's 'Introduction to Psychology,'" *American Journal of Psychology* 50, no. 1/4 (Nov. 1937): 33–36.

24. Ash, "Academic Politics in the History of Science," 273, 283.

25. Alexandra Hui, *The Psychophysical Ear: Musical Experiments, Experimental Sounds, 1840–1910* (Cambridge, MA: MIT Press, 2013), 132. Hui's project examines very different stakes than mine; it is an excellent resource on the intersection of philosophy and psychoacoustics in German experimental psychology.

26. Carl Stumpf and E. M. von Hornbostel, "Über die Bedeutung ethnologischer Untersuchungen für die Psychologie und Ästhetik der Tonkunst," *Beiträge zur Akustik und Musikwissenschaft* 6 (1911): 102.

27. Sebastian Klotz, "Hornbostel, Erich Moritz," in *Die Musik in Geschichte und Gegenwart: allgemeine Enzyklopädie der Musik*, ed. Ludwig Finscher (Kassel: Bärenreiter, 2003): 356.

28. Erich Moritz von Hornbostel, "Die Probleme der vergleichenden Musikwissenschaft" (1905), in *Tonart und Ethos: Aufsätze zur Musikethnologie und Musikpsychologie*, ed. Christian Kaden and Erich Stockmann (Leipzig: Reclam, 1986), 40; cited in Ames, "The Sound of Evolution," 297. Ames offers a much-needed analysis of the connections between Saartjie Baartman's dissection, phonography, and comparative musicology.

29. See New York Public Library Max Wertheimer Papers, Box 11, Series XIII (E. M. von Hornbostel Papers, Bibliographic Notes).

30. Erich Moritz von Hornbostel, "Musikpscyhologische Bemerken über Vogelgesang" [1910], in Kaden and Stockmann, eds., *Tonart und Ethos*, 86–102.

31. Béla Bartók, "Why and How Do We Collect Folk Music?" in *Béla Bartók, Essays*, ed. Benjamin Suchoff (Lincoln, NE: Bison, 1992), 12.

32. Hui, *The Psychophysical Ear*, 143.

33. A copy of the insitute's recording can be found in *The Demonstration Collection of E. M. von Hornbostel and the Berlin Phonogramm-Archiv* (New York: Ethnic Folkways Library, 1963).

34. See, for example, Joseph Jastrow, "Experimental Psychology in Leipzig," *Science* 8, no. 198 (Nov. 1886): 459–62; William Krohn, "Facilities in Experimental Psychology at the Various German Universities," *American Journal of Psychology* 4, no. 4 (Aug. 1892): 585–94; see also Edmund C. Sanford, "A Laboratory Course in Physiological Psychology. (Third Paper): V. Vision," *American Journal of Psychology* 4, no. 3 (Apr. 1892): 474–90.

35. William Krohn, "Facilities in Experimental Psychology at the Various German Universities," *American Journal of Psychology* 4, no. 4 (Aug. 1892): 591.

36. Hermann L. F. Helmholtz, *On the Sensations of Tone as a Physiological Basis for the Theory of Music*, trans. Alexander J. Ellis (London: Longman, Green, [1863] 1875), 30.

37. Edouard-Léon Scott, *The Phonographic Manuscripts of Edouard-Léon Scott de Martinville*, ed. and trans. Patrick Feaster (Bloomington, IN: First Sounds, 2009), 8–9.

38. David Pantalony, *Altered Sensations: Rudolph Koenig's Acoustical Workshop in Nineteenth-Century Paris* (New York: Springer, 2009), 42.

39. Historians have traced Edison's machine to the phonautograph in, for example, Jonathan Sterne, *The Audible Past* (Durham, NC: Duke University Press, 2003), 31–85; Julia Kursell, "A Gray Box: The Phonograph in Laboratory Experiments and Fieldwork," in *The Oxford Handbook of Sound Studies*, ed. Karin Bijsterveld and Trevor Pinch (New York: Oxford University Press, 2012), 176–97. Their stories start with Scott's phonautograph, so in this chapter I am also tracing that history one step farther back into the laboratory tools of vivisection.

40. "Unbildlich: wir brauchen Noten, die freilich durch das Auge wieder hörbar werden, aber auch Zeichen und Schemata, die es nicht werden. Und wenn sich so dem und dem analysierenden Verstande wunderbare Gesetze und Regeln enthüllen, die das Ganze ordnen und noch die unscheinbarste Einzelheit, wie ist es denkbar, daß der Schöpfer ohne Schrift sie gefunden habe, ja, sich ihrer zweifellos nicht mehr bewußt war, als der Spieler und Hörer?" Hornbostel, "Formanalysen an siamesischen Orchesterstücken," 320.

41. In his article, Hornbostel reports identifying key notes in the recording through careful listening; his papers suggest that this type of work was part of standard procedure at the Phonogramm-Archiv. Hornbostel, "Formanalysen an siamesischen Orchesterstücken"; see also correspondence folder, Erich Moritz von Hornbostel Papers, Box 11, Max Wertheimer Collection, New York Public Library Special Collections.

42. Otto Abraham and Erich M. von Hornbostel, "Suggested Methods for the Transcription of Exotic Music," trans. George List and Eve List, *Ethnomusicology* 38, no. 3 (Autumn 1994): 425–56.

43. Jaap Kunst, "A Musicological Argument for Cultural Relationship between Indonesia: Probably the Isle of Java, and Central Africa," *Proceedings of the Musical Association* 62 (1936): 57–76.

44. Alexander J. Ellis, "On the Musical Scales of Various Nations," *Journal of the Society for the Arts* 33, no. 1688 (Mar. 27, 1885): 492–500.

45. See Correspondence Regarding the Wood Pewee, New York State Archives, Box A 3157-55.

46. Albert Brand, "Why Bird Song Can Not Be Described Adequately," *Wilson Bulletin* 49, no. 1 (Mar. 1937): 11–14.

47. Wallace Craig to Albert Brand, April 24, 1935, Correspondence Regarding the Wood Pewee, New York State Archives, Box A 3157–55.

48. Sydney E. Ingraham, "Instinctive Music," *Auk* 55, no. 4 (Oct. 1938): 619.

49. See D. C. Miller, *Science of Musical Sounds* (New York: Macmillan, 1916).

50. See Glen Merry, "Voice Inflection in Speech," *Psychological Monographs of the American Psychological Association*, vol. 31 (Princeton, NJ: Psychological Review, 1922), 205–29; and Max Schoen, "An Experimental Study of the Pitch Factor in Artistic Singing," *Psychological Monographs of the American Psychological Association*, vol. 31 (Princeton, NJ: Psychological Review, 1922), 230–59.

51. Carl E. Seashore, "Phonophotography in the Measurement of the Expression of Emotion in Music and Speech," *Scientific Monthly* 24, no. 5 (May 1927): 463–71; see also Milton Metfessel, *Phonophotography in Folk Music: Negro Songs in New Notation* (Chapel Hill: University of North Carolina Press, 1928); and see Ray E. Miller, "A Strobophotographic Analysis of a Tlingit Indian's Speech," *International Journal of American Linguistics* 6, no. 1 (Mar. 1930): 47–68.

52. Abraham and Hornbostel, "Suggested Methods for the Transcription of Exotic Music," 429.

53. Woodruff Smith, *Politics and the Sciences of Culture in Germany, 1840–1920* (New York: Oxford University Press, 1991).

54. Ash, *Gestalt Psychology in German Culture 1890–1967*; Anne Harrington, *Reenchanted Science: Holism in German Culture from Wilhelm II to Hitler* (Princeton, NJ: Princeton University Press, 1996); Hui, *The Psychophysical Ear*. See also Alexander Rehding, "The Quest for the Origins of Music in Germany Circa 1900," *Journal of the American Musicological Society* 53, no. 2 (Summer 2000): 345–85.

55. Kathleen Kete, "Animals and Ideology" in *Representing Animals*, ed. Nigel Rothfels (Bloomington: Indiana University Press, 2002), 29–30.

56. Wolfgang Köhler, "Gespräche in Deutschland," trans. Ann M. Hentschel, in *Physics and National Socialism: An Anthology of Primary Sources*, ed. Klaus Hentschel (Germany: Birkhäuser, 1996), 36–40.

57. Kunst, "Zum Tode Erich von Hornbostel's," 239–46.

### FIVE  *Postmodern Humanity*

1. See "Art Treasures of Medieval Yugoslavia," *Unesco Courier* 3, no. 4 (May 1950): 4; inset caption, *Unesco Courier* 3, no. 5 (June 1950): 5; Educational Clearing House, "For Ram Chandra, 'Abstracts' are Practical," *Unesco Courier* 3, no. 4 (May 1950): 11; inset caption, *Unesco Courier* 3, no. 4 (May 1950): 9; "U.N. Technical Assistance Mission Surveys Needs of Indonesia," *Unesco Courier* 3, no. 5 (June 1950): 16; cartoon, *Unesco Courier* 3, no. 4 (May 1950): 6; Ritchie Calder, "Men against the Desert," *Unesco Courier* 3, no. 4 (May 1950): 8.

2. "Fallacies of Racism Exposed," *Unesco Courier* 3, no. 6 (July 1950): 1. See also "The Scientific Basis for Human Unity," *Unesco Courier* 3, no. 6 (July 1950): 8–9.

3. Theodosius Dobzhansky, "Race and Humanity," *Science*, new series 113, no. 2932 (Mar. 9, 1951): 264–65; A. H. Sturvetant, "Social Implications of the Genetics of Man," *Science* 120, no. 3115 (Sept. 1954): 405–7; Ernst Mayr, "Review: Origin of the Human Races," *Science*, new series 138, no. 3538 (Oct. 19, 1962): 422; Julian Steward, *Theory of Culture Change: The Methodology of Multilinear Evolution* (Champaign: University of Illinois Press, 1955); Leslie White, *The Evolution of Culture: The Development of Civilization to the Fall of Rome* (Walnut Creek, CA: Left Coast Press, [1959] 2007).

4. Ernst Hans Gombrich, *Art and Illusion* (Princeton, NJ: Princeton University Press, [1960] 2000), 22; see also Thomas Munro, *Evolution in the Arts, and Other Theories of Culture History* (New York: Abrams, 1963).

5. See Warren Dwight Allen, *Philosophies of Music History* (New York: American Book, 1939); William Austin, "The Idea of Evolution in the Music of the Twentieth Century," *Musical Quarterly* 39, no. 1 (Jan. 1953): 26–36; see also Alan Lomax, "Folk Song Style: Notes on a Systematic Approach to the Study of Folk Song," *Journal of the International Folk Music Council* 8 (1956): 48–50.

6. Oscar Riddle, "Genetics and Human Behavior," *American Biology Teacher* 10, no. 3 (Mar. 1948): 69–74.

7. The notion of cultural relativism or comparable trends is reflected in, for example, the loss of historicity in Fredric Jameson's *Postmodernism, or, The Cultural Logic of Late Capitalism* (Durham, NC: Duke University Press, 1991); or the loss of authentic knowledge in Jean-François Lyotard's *The Postmodern Condition: A Report on Knowledge*, trans. Geoff Bennington and Brian Massumi (Minneapolis: University of Minnesota Press, 1984).

8. The term "evolutionary synthesis" was coined first by Julian Huxley in *Evolution: The Modern Synthesis* (London: Allen and Unwin, 1942). Ernst Mayr and others have made the claim that, in addition to synthesizing evolutionary theories of development, this process triggered a secondary "synthesis" that unified biology itself as a discipline. In this chapter, I deal primarily with the ways that the topic of race divided anthropology from biology, rather than with the synthesis itself, which deserves much more complex mention. Mayr and Provine wrote the dominant history of the synthesis in their preface to *The Evolutionary Synthesis: Perspectives on the Unification of Biology*, ed. Ernst Mayr and William B. Provine (Cambridge, MA: Harvard University Press, 1980), ix.

9. See Mayr and Provine, eds., *The Evolutionary Synthesis*; see also Vassiliki Betty Smocovitis, *The Evolutionary Synthesis and Evolutionary Biology* (Princeton, NJ: Princeton University Press, 1996).

10. Jenny Reardon, *Race to the Finish: Identity and Governance in an Age of Genomics* (Princeton, NJ: Princeton University Press, 2005), 75.

11. Donald Jay Grout, *A Short History of Opera*, vol. 1 (New York: Columbia Univer-

sity Press, 1947), 9. See also Donald Jay Grout, *A History of Western Music* (New York: Norton, 1960), xi.

12. In this section, I reprise a discussion of "the music itself" that is explored in more depth in "Musical Style from Adler to America," *Journal of the American Musicological Society* 67, no. 3 (Fall 2014): 735–68.

13. Daniel Gregory Mason, *The Art of Music*, vol. 1 (New York: National Society of Music, 1915). See also Jules Comabrieu, *La Musique: ses Lois, son* Évolution (Paris: Flammarion, 1907); Charles Hubert Parry, *The Evolution of the Art of Music*, ed. H. C. Colles (New York: Appleton, [1893] 1921); and Edward Dickinson, *The Study of the History of Music* (New York: Scribner), 1920.

14. Vincent D'Indy and Auguste Sérieyx, *Cours de Composition Musicale*, vol. 1 (Paris: Durand, 1912), 77. For an overview of D'Indy's antisemitism and its influence, see Jane Fulcher's "Vincent d'Indy's 'Frame Anti-Juif' and Its Meaning in Paris, 1920," *Cambridge Opera Journal* 2, no. 3 (Nov. 1990): 295–319.

15. Richard Crocker, *A History of Musical Style* (New York: Dover, 1966, 1986), vi.

16. Grout, *A Short History of Opera*, vii.

17. Ibid., 386–87.

18. Grout, *A History of Western Music*, xi.

19. Albert T. Luper, "Review: A History of Western Music by Donald Jay Grout," *Notes* 18, no. 1 (Dec. 1960): 47; Louise Rood, "An Invaluable Survey of Western Music," *Massachusetts Review* 2, no. 1 (Autumn 1960): 180.

20. Warren Dwight Allen, "Review: A History of Western Music," *Journal of Research in Music Education* 8, no. 2 (Autumn 1960): 125.

21. See, for example, several representative reviews of Grout's book: Luper, "Review: *A History of Western Music* by Donald Jay Grout"; J. A. W., "Review: *A History of Western Music* by Donald Jay Grout," *Music and Letters* 43, no. 4 (Oct. 1962): 363–65; Alec Harman, "Review: One-Volume History," *Musical Times* 109, no. 1438 (Dec. 1962): 845, 847.

22. In addition to Grout's writing, key texts in the movement include Crocker, *A History of Musical Style*; Jan LaRue, *Guidelines for Style Analysis* (New York: Norton, 1970); Charles Rosen, *The Classical Style: Haydn, Mozart, Beethoven* (New York: Norton, 1971).

23. See, for example, Michael Denning, *Noise Uprising: The Audiopolitics of a World Musical Revolution* (New York: Verso, 2015); Timothy D. Taylor, "Radio," in *Music, Sound, and Technology in America: A Documentary History of the Early Phonograph, Cinema, and Radio* (Durham, NC: Duke University Press, 2012), 329–54.

24. Gwynne Kuhner Brown, "Problems of Race and Genre in the Critical Reception of Porgy and Bess" (doctoral diss., University of Washington), 2006. Descriptions of Cold War music propaganda can be found in René Thimonnier, "Principes d'une Propagande musicale française en Allemagne occupé," *La Revue Musicale* 201 (Sept. 1946): 313–14; see also Leslie A. Sprout, "The 1945 Stravinsky Debates: Nigg, Messiaen, and the Early Cold

War in France," *Journal of Musicology* 26, no. 1 (Winter 2009): 85–131; Mark Carroll, *Music and Ideology in Cold War Europe* (New York: Cambridge University Press, 2003); Elizabeth B. Crist, "Aaron Copland and the Popular Front," *Journal of the American Musicological Society* 56, no. 2 (Summer 2002): 409–65.

25. See Bach family tree, Eugenics Record Office Papers, American Philosophical Society Library.

26. Edmund W. Sinnott, *Albert Francis Blakeslee* (Washington, DC: National Academy of Sciences, 1959), 22.

27. Glen Haydon, *Introduction to Musicology: A Survey of the Fields, Systematic and Historical, of Musical Knowledge and Research* (Westport, CT: Greenwood, [1941] 1978), 253.

28. Austin, "The Idea of Evolution," 26, 27.

29. Paul Lang, "Music Scholarship at the Crossroads," *Musical Quarterly* 31, no. 3 (July 1945): 373.

30. See, for example, Austin, "The Idea of Evolution," 35; Alan P. Merriam, *The Anthropology of Music* (Evanston, IL: Northwestern University Press, 1964), 17.

31. Bruno Nettl discusses this terminological shift in "Comparison and Comparative Method in Ethnomusicology," *Anuario Interamericano de Inverstigacion Musical* 9 (1973): 148.

32. Alan Lomax and Norman Berkowitz, "The Evolutionary Taxonomy of Culture," *Science* 177, no. 4045 (July 21, 1972): 228.

33. "Il ne faisait de doute pour personne que la perfection—celle de l'art autant que toute autre—est l'aboutissement des incessants progrès accomplis par l'humanité, grâce au génie particulier des 'Blancs' et que, par voie de consequence, seuls ces Blancs possèdent une musique parfaite, mesure de toute chose musicale." Originally published in 1960, in Constantin Braïloiu, "La Vie Antérieure," in *Histoire de la Musique: des origines à Jean-Sébastien Bach*, vol. 1, ed. Alexis Roland-Manuel (Paris: Éditions Gallimard, 1986), 119. See also Alexis Roland-Manuel's preface to the same volume.

34. Curt Sachs, *The Wellsprings of Music* (New York: Da Capo, 1962), 47–48.

35. Ibid., 48.

36. Ibid. Intriguingly, Sachs laid this confusion between genetics and culture at the feet of François-Joseph Fétis, a French music scholar whose racial categories were, in part, remedied by the Sachs-Hornbostel instrument classification system.

37. Amos G. Avery, Sophie Satina, and Jacob Rietsema; "Foreword and Biographical Sketch by Edmund W. Sinnott," in *Blakeslee: The Genus Datura* (New York: Ronald Press, 1959), 19.

38. Samuel N. F. Sanford, *New England Herbs, Their Preparation and Use* (Boston: New England Museum of Natural History, 1937), 18–19; Thomas H. Kearney and Robert H. Peebles, *Flowering Plants and Flowers of Arizona* (Washington, DC: United States Department of Agriculture, 1942), 797.

39. Edmund Sinnott recounts this story and quotes Blakeslee's letter in Sinnott, *Albert Francis Blakeslee*, 9.

40. For a fuller telling of this story, see Nathaniel Comfort's *The Science of Human Perfection: How Genes Became the Heart of American Medicine* (New Haven: Yale University Press, 2012). See also Bentley Glass, "Milislav Demerec 1895–1966," (Washington, DC: National Academy of Sciences, 1971).

41. Comfort, *The Science of Human Perfection*, 69–70.

42. Nathaniel Comfort, *The Tangled Field: Barbara McClintock's Search for the Patterns of Genetic Control* (Cambridge, MA: Harvard University Press, 2003), 81.

43. Harvard professor W. J. Crozier, quoted in Vassiliki Betty Smocovitis, "The Evolutionary Synthesis and Evolutionary Biology," *Journal of the History of Biology* 25, no. 1 (Spring 1992): 16. Smocovitis describes the state of evolutionary biology at the dawn of the synthesis here and elsewhere with considerable depth and breadth.

44. Edward John Larson, *Evolution: The Remarkable History of a Scientific Theory* (New York: Modern Library, 2004), 300.

45. See J. B. S. Haldane, "Differences," *Mind* 57, no. 227 (July 1948): 296; Riddle, "Genetics and Human Behavior," 74; A. H. Sturtevant, "Social Implications of the Genetics of Man," *Science* 120, no. 3115 (Sept. 10, 1954): 405.

46. Albert Blakeslee, "Ideals of Science," *Science* 92, no. 2400 (Dec. 27, 1940): 592.

47. Comfort provides an intriguing overview of early genetics and its ties to eugenics research in *The Science of Human Perfection*.

48. Ibid., 97.

49. Reardon, *Race to the Finish*, 61.

50. Blakeslee didn't specify what, exactly, "wayward" meant, but he made reference to the development of weapons by scientists. In 1940, he might have been responding to rumors of the Holocaust or simply the broader criticism that science had made war more lethal. Blakeslee, "Ideals of Science," 589.

51. James F. Crow, "Birth Defects, Jimson Weeds and Bell Curves," *Genetics* 147 (Sept. 1997): 2; Sinnott, *Albert Francis Blakeslee*, 22.

52. Laura Craytor Boulton describes those tents in a letter to Charles Davenport, March 27, 1922, Charles Davenport Papers Subseries B. Box 110, Laura Boulton Folder, American Philosophical Society Library.

53. Crow, "Birth Defects, Jimson Weeds and Bell Curves," 2.

54. See Edwin Teale, "Test-tube Magic Creates Amazing New Flowers," *Popular Science* (May 1940): 58.

55. Milislav Demerec, "Foreword," in *Origin and Evolution of Man*, Cold Spring Harbor Symposia on Quantitative Biology XV, ed. Katherine Brehme Warren (Cold Spring Harbor, NY: Biological Laboratory at Cold Spring Harbor, 1951), v.

56. Ibid.

57. Smocovitis quotes Ernst Mayr as noting Washburn's paper as a turning point in the relation of physical anthropology to the new evolutionary synthesis. Vassilliki Betty Smocovitis, "Humanizing Evolution: Anthropology, the Evolutionary Synthesis, and the Prehistory of Biological Anthropology, 1927–1962," *Current Anthropology* 53, no. S5 (Apr. 2012): S115–116.

58. Smocovitis, "Humanizing Evolution"; Reardon, *Race to the Finish*, 61.

59. Smocovitis has discussed this topic at length in her excellent history of the impact of the Darwin centennial in 1959 on the synthesists, "'It Ain't Over 'Til It's Over': Rethinking the Darwinian Revolution," *Journal of the History of Biology* 38, no. 1 (Spring 2005): 33–49.

60. Smocovitis, *The Evolutionary Synthesis and Evolutionary Biology*.

61. Ronald Amundson discusses the debates about typology, population genetics, and racial stereotyping in *The Changing Role of the Embryo in Evolutionary Thought* (New York: Cambridge University Press, 2005), 204–6.

62. Ernst Mayr, "Review: Origin of the Human Races," *Science*, new series 138, no. 3538 (Oct. 19, 1962): 422.

63. Theodosius Dobzhansky, "Race and Humanity," *Science*, new series 113, no. 2932 (Mar. 9, 1951): 264–65.

64. Ibid., 264.

65. For a thorough and thoughtful overview of anthropology's response to postwar changes in evolutionary theory, see Timothy Ingold, *Evolution and Social Life* (New York: Cambridge University Press, 1986).

66. Margaret Mead, *Continuities in Cultural Evolution* (New Brunswick, NJ: Transaction, [1964] 1999), 5.

67. Julian Steward explained, "In biological evolution it is assumed that all forms are genetically related and that their development is essentially divergent . . . In cultural evolution, on the other hand, it is assumed that cultural patterns in different parts of the world are genetically unrelated." Julian Steward, *Theory of Culture Change: The Methodology of Multilinear Evolution* (Champaign: University of Illinois Press, [1955] 1972), 12.

68. Morris Ginsberg, "Social Evolution," in *Darwinism and the Study of Society*, ed. Michael Banton (Chicago: Quadrangle, 1961), 101–2. See also, for example, Alexander Alland's distinction between "real" biology and anthropology in *Evolution and Human Behavior* (New York: Natural History Press, 1967), 194.

69. Ernest Beaglehole et al., "Statement on Race: 1950," in *Four Statements on the Race Question* (Paris: United Nations, 1969), 30–35.

70. Beaglehole et al., "Statement on Race: 1950," 30.

71. Ibid., 31.

72. The concept of doublethink, which I use here, was actually described years after the UNESCO statement by Orwell: "holding two contradictory beliefs in one's mind simultaneously, and accepting both of them." George Orwell, *Nineteen Eighty-Four* (New

York: Signet Classics, 1977), 215. Alternately, Orwell describes doublethink this way: "To know and not to know, to be conscious of complete truthfulness while telling carefully constructed lies, to hold simultaneously two opinions which cancelled out, knowing them to be contradictory and believing in both of them, to use logic against logic, to repudiate morality while laying claim to it, to believe that democracy was impossible and that the Party was the guardian of democracy, to forget whatever it was necessary to forget, then to draw it back into memory at a time when it was needed, and then promptly to forget it again, and above all, to apply the same process to the process itself—that was the ultimate subtlety: consciously to induce unconsciousness, and then, once again, to become unconscious of the act of hypnosis you had just performed. Even to understand the word "doublethink" involved the use of doublethink." Ibid, 35.

73. See, for example, Albert Blakeslee, "Corn and Education," *Journal of Education* 88, no. 12 (Oct. 3, 1918): 326.

74. See, for example, Eric Voegelin, "The Growth of the Race Idea," *Review of Politics* 2, no. 3 (July 1940): 283; William C. Boyd, "The Use of Genetically Determined Characters, Especially Serological Factors Such as Rh, in Physical Anthropology," *Southwestern Journal of Anthropology* 3, no. 1 (Spring 1947): 32.

75. See Blakeslee, "Ideals of Science."

76. Comfort, *The Science of Human Perfection*, 114.

77. Wilton Marion Krogman, "Physical Anthropology and Race Relations: A Biosocial Evaluation," *Scientific Monthly* 66, no. 4 (Apr. 1948): 321.

78. Ibid., 320–21.

79. Ibid., 319.

80. At the very beginning of World War II, Blakeslee wrote that although science was impersonal, politics had "bombed" science as effectively as the war was flattening buildings, describing how the announcement of war in Edinburgh had broken up the 1939 International Congress of Genetics, forcing attendees from Axis countries to leave the meeting. Blakeslee, "Ideals of Science," 592.

81. See Rachel Mundy, "Evolutionary Categories and Music Style from Adler to America," *Journal of the American Musicological Society* 67, no. 3 (Fall 2014): 761; Alexander Rehding, "The Quest for the Origins of Music in Germany Circa 1900," *Journal of the American Musicological Society* 53, no. 2 (Summer 2000): 345–85; Bennett Zon, *Representing Non-Western Music in Nineteenth-Century Britain* (Rochester, NY: University of Rochester Press, 2007); Benjamin Steege, *Helmholtz and the Modern Listener* (New York: Cambridge University Press, 2012). See also Mundy, "The 'League of Jewish Composers' and American Music," *Musical Quarterly* 96, no. 1 (Spring 2013): 50–99.

82. See Cold Spring Harbor Laboratory's Image Archive on the American Eugenics Movement, eugenicsarchive.org.

83. Comfort has described the recasting of Davenport in *The Science of Human Perfection*, 240; one can also see elements of this counter-history in the online history of Cold Spring Harbor (https://www.cshl.edu/About-Us/History.html, accessed Sept. 22, 2015), in which the presence and importance of the Eugenics Record Office is literally not present, replaced instead with a section titled "1920s–30s Cornerstones of Modern Cancer Research."

### SIX  *Listening for Objectivity*

1. I am grateful to David Novak for inspiring this comparison between the spectrograph and the melograph, by sharing his thoughts on the subject of mechanical transcription with me in 2010.

2. Robert A. Hinde, "William Homan Thorpe," *Biographical Memoirs of Fellows of the Royal Society* 33 (Dec. 1987): 627.

3. Ann M. Pescatello, *Charles Seeger: A Life in American Music* (Pittsburgh: University of Pittsburgh Press, 1992), 202–12.

4. Theodor Adorno, "On the Fetish Character in Music and the Regression of Listening" [1938] in Adorno, *The Culture Industry: Selected Essays on Mass Culture*, ed. J. M. Bernstein (New York: Routledge, 1991), 29–60; Pierre Bourdieu, "The Aristocracy of Culture," trans. Richard Nice, *Media Culture Society* 2 (1980): 225–54.

5. Mitchell Ash, *Gestalt Psychology in German Culture 1890–1967: Holism and the Quest for Objectivity* (New York: Cambridge University Press, 1998), 30–31, 94, 114–15, 190.

6. See Mark Carroll, *Music and Ideology in Cold War Europe* (New York: Cambridge University Press, 2003); Toby Thacker, *Music After Hitler, 1945–1955* (Burlington, VT: Ashgate, 2007); René Thimonnier, "Principes d'une Propagande musicale française en Allemagne occupé," *La Revue Musicale* 201 (Sept. 1946): 313–14.

7. See Jon Gertner, *The Idea Factory: Bell Labs and the Great Age of American Innovation* (New York: Penguin, 2012), 59–74.

8. Ralph Potter, "Visible Patterns of Sound," *Science* 102, no. 2654 (Nov. 9, 1945): 463–70; Cody J. Ham, "How Does Your Voice Look Sent Over the Telephone?," *Telephony* 64 (Jan.–July 1913): 92.

9. Otto Abraham and Erich M. von Hornbostel, "Suggested Methods for the Transcription of Exotic Music," trans. George List and Eve List, *Ethnomusicology* 38, no. 3 (Autumn 1994): 429.

10. Lorraine Daston and Peter Galison, *Objectivity* (New York: Zone, 2010).

11. This history and its variants are told in many sources. Ornithological accounts include Luis F. Baptista and Sandra L. Gaunt, "Advances in Studies of Avian Sound Communication," *Condor* 96, no. 3 (Aug. 1994): 817–30; Peter Slater, "Fifty Years of Bird Song

Research: A Case Study in Animal Behavior," *Animal Behavior* 65, no. 4 (2003): 633–39; Peter Marler and Hans Slabbekoorn, *Nature's Music: The Science of Birdsong* (New York: Elsevier, 2004). Gregory Radick offers a particularly rich perspective informed by prewar attempts to study animal voices. Gregory Radick, *The Simian Tongue: The Long Debate about Animal Language* (Chicago: University of Chicago Press, 2007).

12. Radick, *The Simian Tongue*; see also William H. Thorpe, "The Process of Song-Learning in the Chaffinch as Studied by Means of the Sound Spectrograph," *Nature* 173, no. 4402 (Mar. 13, 1954): 465.

13. Baptista and Gaunt, "Advances in Studies of Avian Sound Communication," 818; Marler and Slabbekoorn, *Nature's Music*, 1.

14. Radick, *The Simian Tongue*, 265.

15. Thorpe, "The Process of Song-Learning in the Chaffinch," 465.

16. Ibid., 465–66.

17. See William H. Thorpe, "Some Implications of the Study of Animal Behavior," *Scientific Monthly* 84, no. 6 (June 1957): 309–20; William H. Thorpe, "Comparative Psychology," *Annual Review of Psychology* 12 (1961): 27–50.

18. Thorpe, "Some Implications of the Study of Animal Behavior," 309.

19. Ibid.

20. Ibid., 309, 310, 311. See also Charles E. Spearman, "The Confusion That Is Gestalt-Psychology," *American Journal of Psychology* 50, no. 1 (Nov. 1937): 381.

21. Thorpe cites Craig in several places, including in William H. Thorpe, "The Learning of Song Patterns by Birds, with Especial Reference to the Song of the Chaffinch Fingilla Coelebs," *Ibis* 100, no. 4 (1958): 535–70. Marler argued that Craig's teacher Otis Whitman should be treated as a founding figure in ethology (Peter Marler, "Ethology and the Origins of Behavioral Endocrinology," *Hormones and Behavior* 47 [2005]: 493–502), and described Craig in a 2004 interview with David Rothenberg as a "pioneer." Peter Marler and David Rothenberg, "Interview with Peter Marler," Terrain.org (Dec. 8, 2014), accessed Aug. 12, 2015, http://www.terrain.org/2014/interviews/peter-marler/.

22. Thorpe, "The Process of Song-Learning in the Chaffinch," 465-69.

23. See, for example, Ibid., 466.

24. Ibid., 465.

25. Hinde, "William Homan Thorpe," 630.

26. Ibid.

27. See William H. Thorpe, "The Biological Significance of Duetting and Antiphonal Song," *Acta Neurobiologiae Experimentalis* 35, no. 5–6 (1975): 518, 519, 520.

28. Baptista and Gaunt, "Advances in Studies of Avian Sound Communication," 818.

29. Marler and Slabbekoorn, *Nature's Music*, 1.

30. Joeri Bruyninckx's invaluable history of this transition notes that figures such as Thorpe engaged in debates about whether to stop using musical notation in studies of

birdsong, though he says the controversy was not a significant one. Joeri Bruyninckx, "Sound Science: Recording and Listening in the Biology of Bird Song, 1880–1980" (PhD diss. University of Maastricht, 2012), 142–44.

31. Otto Abraham and Erich von Hornbostel, "Vorschläge zur Transkription exotischer Melodien," *Sammelbände der Internationalen Musikgsellschaft* 11, no. 1 (1909): 1–25.

32. Charles Seeger, "An Instantaneous Music Notator," *Journal of the International Folk Music Council* 3 (1951): 103.

33. Béla Bartók, *Serbo-Croatian Folk Songs*, ed. and trans. Albert B. Lord (New York: Columbia University Studies in Musicology, 1951), 3.

34. Charles Seeger, "Toward a Universal Music Sound-Writing for Musicology," *Journal of the International Folk Music Council* 9 (1957): 66.

35. Seeger called the distinction "prescriptive" (instructions on how to perform a song) and "descriptive" (an empirical description of the performance), a pair of terms that are often still circulated in ethnomusicology. Charles Seeger, "Prescriptive and Descriptive Music-Writing," *Musical Quarterly* 44, no. 2 (Apr. 1958): 184–95.

36. Seeger, "An Instantaneous Music Notator," 106.

37. Pescatello, *Charles Seeger*, 213.

38. Jaap Kunst, "Zum Tode Erich von Hornbostel's," *Anthropos* 32, no. 1/2 (Jan.–Apr. 1937): 246; George List and Donald Gray, "Memorial Resolution: George Harold List," Circular B16 2009, Indiana University (2009), accessed Aug, 12, 2015, http://www.indiana .edu/~bfc/docs/circulars/08–09/B16–2009.pdf.

39. George List, *Stability and Variation in Hopi Song* (Philadelphia: American Philosophical Society, 1993).

40. Seeger, "An Instantaneous Music Notator," 106.

41. Seeger, "Prescriptive and Descriptive Music-Writing," 185, 194.

42. Ibid., 185, 186, 187.

43. C. T. Onions, ed., *The Oxford Dictionary of English Etymology* (New York: Oxford University Press, 1966), 620.

44. Daston and Galison, *Objectivity*.

45. Those who complained of bafflement included Frederick Batchelder, "The Transcription of Bird Songs," *Primary Education* 12, no. 3 (May 1904): 217; Percy Grainger, "Collecting with the Phonograph," *Journal of the Folk-Song Society* 3, no. 12 (1908): 148; Aretas Saunders, "Some Suggestions for Better Methods of Recording and Studying Bird Songs," *Auk* 32, no. 2 (Apr. 1915): 174; Percy Grainger, "The Impress of Personality in Unwritten Music," *Musical Quarterly* 1, no. 3 (July 1915): 422.

46. Grainger, "Collecting with the Phonograph," 152.

47. See Charles K. Wead, "The Study of Primitive Music," *American Anthropologist* 2, no. 1 (Jan. 1900): 79; Eugenie Lineff, *The Peasant Songs of Great Russia*, second series (Moscow: Municipal Printing Office, 1912), 58; Frances Densmore, *Teton Sioux Music*

(Washington, DC: Smithsonian Institution Bureau of American Ethnology Bulletin 61, 1918), 51; Carroll Brent Chilton, *The De-assification of Music: A Propagandist Magazine of One Number, Containing News of Importance to All Music Lovers, Especially to All Owners of Player Pianos* (New York: Chilton, 1922), 9; Ray E. Miller, "A Strobophotographic Analysis of a Tlingit Indian's Speech," *International Journal of American Linguistics* 6, no. 1 (Mar. 1930): 48, 49; Carl E. Seashore, "Psychology in Music: The Rôle of Experimental Psychology in the Science and Art of Music," *Musical Quarterly* 16, no. 2 (Apr. 1930): 235.

48. Erich Moritz von Hornbostel and Otto Abraham equated accuracy with objectivity and framed graphing as tool that increased the clarity of a researcher's overall impression of music. "Suggested Methods for the Transcription of Exotic Music," trans. George List and Eve List, *Ethnomusicology* 38, no. 3 (Autumn, 1994) [1909–10]: 427, 429. Lineff writes of "the objective method, that is to say, . . . graphic representation" (*The Peasant Songs of Great Russia*, 58). Densmore used "object" to refer to her graphic images of songs, rather than to her musical notations (*Teton Sioux Music*, 51). Helen Roberts likewise described graphic scores as revealing the "object" of music, by which she meant its typological characteristics. Helen Roberts, "New Phases in the Study of Primitive Music," *American Anthropologist* 24, no. 2 (Apr.–Jun. 1922): 148. See also Saunders, "Some Suggestions for Better Methods of Recording and Studying Bird Songs"; Sydney E. Ingraham, "Instinctive Music," *Auk* 55, no. 4 (Oct. 1938): 619.

49. Chilton, *The De-assification of Music.*

50. Frederic Lees, "Photographing Speech," photographs by Laurence, Paris, *Windsor Magazine* 27 (London: Ward, Lock, Jan. 1908): 257–64.

51. See, for example, Miller, "A Strobophotographic Analysis of a Tlingit Indian's Speech," 48, 49; Seashore, "Psychology in Music," 235; Albert Brand, "A Method for the Intensive Study of Bird Song," *Auk* 52, no. 1 (Jan. 1935): 40–52.

52. Charles Spearman, one of Gestalt's critics, claimed that the Gestalt school originated with the question of how human listeners heard a melody, instead of discrete but separate tones. Spearman, "The Confusion That Is Gestalt-Psychology," 370.

53. Spearman surveyed these critiques of the Gestalt in 1937, citing examples in Italian, German, and English. Ibid.

54. Paul R. Farnsworth, "Sacred Cows in the Psychology of Music," *Journal of Aesthetics and Art Criticism* 7, no. 1 (Sept. 1948): 49; A. W. Wolters, "Review: Principles of Gestalt Psychology by K. Koffka," *Philosophy* 11, no. 44 (Oct. 1936): 503; Spearman, "The Confusion That Is Gestalt-Psychology," 381.

55. See Carl E. Seashore, "Phonophotography in the Measurement of the Expression of Emotion in Music and Speech," *Scientific Monthly* 24, no. 5 (May 1927): 463–71; Milton Metfessel, *Phonophotography in Folk Music: American Negro Songs in New Notation*

(Chapel Hill: University of North Carolina Press, 1928); Miller, "A Strobophotographic Analysis of a Tlingit Indian's Speech"; Milton Metfessel, "Strobophotography in Bird Singing," *Science* 79, no. 2053 (May 4, 1934): 412–13.

56. Glen Merry, "Voice Inflection in Speech," *Psychological Monographs of the American Psychological Association*, vol. 31 (Princeton, NJ: Psychological Review, 1922), 207.

57. Metfessel, "The Collecting of Folk Songs by Phonophotography," 28.

58. Albert Brand, "Why Bird Song Can Not Be Described Adequately," *Wilson Bulletin* 49, no. 1 (Mar. 1937): 12.

59. One of the early reviews of Koffka's work, for example, complained that it was "pure deduction from first principles." Wolters, "Review: Principles of Gestalt Psychology by K. Koffka," 503. And Paul Farnsworth suggested that the idea that aesthetics could be studied rigorously using Gestalt principles was a "sacred cow." Farnsworth, "Sacred Cows in the Psychology of Music," 49.

60. Seeger, "An Instantaneous Music Notator," 103.

61. Joseph Campbell, *The Hero with a Thousand Faces*, 3rd ed. (Novato, CA: New World Library, 2008), 11.

62. Baptista and Gaunt, "Advances in Studies of Avian Sound Communication," 818; Marler and Slabbekoorn, *Nature's Music*, 1.

63. Donald Kroodsma, *The Singing Life of Birds* (New York: Houghton Mifflin, 2005), 1.

64. L. Irby Davis, "Biological Acoustics and the Use of the Sound Spectrograph," *Southwestern Naturalist* 9, no. 3 (Aug. 20, 1964): 119–20.

65. Peter Marler explained in 2005, "The original spectrograph produced images that were very hard to reproduce, so I came up with this method of tracing them. Later came the more Japanese ink-painting filled-in approach. This was worked out with . . . [the] editor of *Ibis*, he could not handle sonograms, so we had to work out a way to present them in the middle 50s." Marler and Rothenberg, "Interview with Peter Marler."

66. Baptista and Gaunt, "Advances in Studies of Avian Sound Communication," 818.

67. Craig Eley has suggested that a similar preference for short, repetitive birdsong was the norm for professional whistlers, who predated the spectrograph by several decades. Craig Eley, "A Birdlike Act: Sound Recording, Nature Imitation, and Performance Whistling," *Velvet Light Trap* 74 (Fall 2014): 1–15.

68. One of the more intriguing exceptions to this rule is Thorpe's later work on duetting in birds, a complex behavior that unfolds over longer time spans. He used musical notation in this work, and described songs in the terms of older comparative psychologists like Craig. Yet it is Thorpe's chaffinch research that has entered the canon of ornithology. William Homan Thorpe, *Duetting and Antiphonal Song in Birds: Its Extent and Significance* (Leiden, Germany: Brill), 1972; Thorpe, "The Biological Significance of Duetting and Antiphonal Song."

69. Donald Kroodsma and Noah K. Strycker, "A *Birding* Interview with Donald Kroodsma," *Birding* 41, no. 3 (May 2009): 20.

70. Marler and Rothenberg, "Interview with Peter Marler."

71. Pescatello, *Charles Seeger*, 213.

72. George List, "The Musical Significance of Transcription (Comments on Hood, 'Musical Significance')," *Ethnomusicology* 7, no. 3 (Sept. 1963): 194–95.

73. George List, "The Reliability of Transcription," *Ethnomusicology* 18, no. 3 (Sept. 1974): 376.

74. Nazir A. Jairazbhoy, "The 'Objective' and Subjective View in Music Transcription," *Ethnomusicology* 21, no. 2 (May 1977): 264.

75. Ibid.

76. Timothy Rice, *Ethnomusicology: A Very Short Introduction* (New York: Oxford University Press, 2014), 22; Gregory F. Barz and Timothy J. Cooley, *Shadows in the Field: New Perspectives for Fieldwork in Ethnomusicology*, 2nd. ed. (New York: Oxford University Press, 2008), 25; Bruno Nettl, *The Study of Ethnomusicology: Thirty-One Issues and Concepts* (Champaign: University of Illinois Press, 2005), 91.

77. There were a few notable exceptions on both sides: in the early 1980s, linguist Ray Jackendoff worked with composer Fred Lerdahl to apply Chomskyian linguistics to the problem of human music's relation to birdsong, while Alan Lomax's cantometrics project served as an ongoing counterpart to critiques of music science like List's and Jairazbhoy's during the 1970s. Ray Jackendoff and Fred Lerdahl, *A Generative Theory of Tonal Music* (Cambridge, MA: MIT Press, 1981).

78. Barz and Cooley, *Shadows in the Field*, 25.

79. Theodora J. Kalikow and John A. Mills, "Wallace Craig (1876–1954), Ethologist and Animal Psychologist," *Journal of Comparative Psychology* 103, no. 3 (1989): 281.

80. Wallace Craig to Leonard Carmichael, July 14, 1944, Box 1, Craig correspondence, Leonard Carmichael Papers, American Philosophical Society Library.

81. Report submitted by Wallace Craig to Leonard Carmichael, page 4, Oct. 10, 1944, Box 1, Craig correspondence, Leonard Carmichael Papers, American Philosophical Society Library.

82. Wallace Craig, *The Song of the Wood Pewee Myochanes Virens Linnaeus: A Study of Bird Music* (Albany: New York State Museum Bulletin no. 334, June 1943), 156.

83. Ibid., 157.

84. Köhler to Leonard Carmichael, May 30, 1944, Box 1, Leonard Carmichael Papers, American Philosophical Society Library.

85. Wallace Craig to Leonard Carmichael, Oct. 10, 1944, Box 1, Craig correspondence, Leonard Carmichael Papers, American Philosophical Society Library.

1. The popularity of audio bird guides fares particularly well in comparison to guides to musical style. On Amazon.com (accessed Apr. 27, 2010), Peterson's *Birding by Ear: Eastern and Central North America* (Boston: Houghton Mifflin, 2002) ranked at 6,188—amazingly, it was number 4 in Amazon's nonfiction books on CD in 2010. Compare this to *The Rough Guide to World Music: Africa and Middle East* (London: World Music Network, 2007), which ranked at 80,487, around thirteen times less popular (classical music guides were even less popular, and a surprising number of them with the word "guide" in the title were designed to help brides pick wedding music). By 2016, birders were more likely to use smartphone apps or websites; but the Peterson guide still ranked well, at number 11 in nonfiction books on CD and 28,671 in books. Its sales continued to compare favorably to music guides such as Titon and Cooley's *Worlds of Music* ebook, which ranks at 211,387 in the Kindle Store (accessed Nov. 22, 2016).

2. R. Murray Schafer, *The Soundscape: Our Sonic Environment and the Tuning of the World* (Rochester, VT: Destiny, [1977] 1994).

3. Ibid., 246.

4. For example, Kay Kaufman Shelemay's *Soundscapes: Exploring Music in a Changing World* (New York: Norton, 2001), published in the same year as Feld's *Soundwalks*, uses "soundscape" without its environmentalist connotations.

5. See Schafer, *Soundscape*; see also Bernie Krause, *Wild Soundscapes: Discovering the Voice of the Natural World* (Berkeley: Wilderness Press, 2002) for an extension of acoustic ecology into contemporary recording experiments.

6. This chapter builds on my essays "Museums of Sound: Audio Bird Guides and the Pleasures of Knowledge," *Sound Studies* 2 (2016), http://dx.doi.org/10.1080/20551940 .2016.1228842; and "Birdsong and the Image of Evolution," *Society and Animals* 17, no. 3 (2009): 206–23.

7. John Law and Michael Lynch, "Field Guides and the Descriptive Organization of Seeing: Birdwatching as an Exemplary Observational Activity," *Human Studies* 11, no. 2 (1988): 271–303; Helen Macdonald, "'What Makes You a Scientist Is the Way You Look at Things': Ornithology and the Observer 1930–1955," *Studies in History and Philosophy of Biological and Biomedical Sciences* 33 (2002): 55–77; Thomas R. Dunlap, "Tom Dunlap on Early Bird Guides," *Environmental History* 10, no. 1 (2005): 110–18; Dunlap, *In the Field, Among the Feathered: A History of Birders and Their Guides* (Oxford: Oxford University Press, 2011).

8. Law and Lynch, "Field Guides and the Descriptive Organization of Seeing"; Linda Dugan Partridge, "By the Book: Audubon and the Tradition of Ornithological Illustration," *Huntington Library Quarterly* 59, no. 2/3 (1996): 269–301.

9. In Eaton's guide, for example, members of the duck family were described this way.

Elon Howard Eaton and Louis Agassiz Fuertes, *Birds of New York* (New York: New York State Museum, 1910), 283–26. See also Dunlap, *In the Field, Among the Feathered*.

10. Eaton and Fuertes, *Birds of New York*, 9.

11. "An intelligent and apparently honest man tells me that he once ate of an owl . . . and found it as sweet and tender fowl as he had ever tasted. So it seems the owl is not always stupid nor always tough." Simeon Pease Cheney, *Wood Notes Wild: Notations of Bird Music* (Boston: Lee and Shepard, 1891), 103.

12. Law and Lynch, "Field Guides and the Descriptive Organization of Seeing," 278–86. See also Roger Tory Peterson, *A Field Guide to the Birds of Eastern and Central North America* (Boston: Houghton Mifflin, 1980); Valérie Chansigaud, *Histoire de l'ornithologie* (Paris: Delachaux et Niestlé, 2007), 192–93.

13. Peterson, *A Field Guide to the Birds of Eastern and Central North America*, 23–28.

14. Ibid., 7.

15. Arthur A. Allen and Peter Paul Kellogg, *American Bird Songs* (Ithaca, NY: Comstock, 1942), 78 rpm.

16. This story is told at length in Joeri Bruyninckx's "Sound Science: Recording and Listening in the Biology of Bird Song 1880–1980" (PhD diss., University of Maastricht, 2012) and "Sounds Sterile: Making Scientific Field Recordings in Ornithology," in *The Oxford Handbook of Sound Studies*, ed. Karin Bijsterveld and Trevor Pinch (New York: Oxford University Press, 2011): 127–50.

17. See Albert Brand's "A Method for the Intensive Study of Bird Song," *Auk* 52, no. 1 (1935): 40–52; *Songs of Wild Birds* (New York: Thomas Nelson, 1934); "Recording Sounds of Wild Birds," *Auk* 49, no. 4 (1932): 436–39.

18. Lang Elliott, Donald Stokes, and Lilian Stokes, *Stokes Field Guide to Birdsongs* (Time Warner Audio Books, 2–32482, 1997).

19. See, for example, the liner notes to the Peterson guide *Backyard Bird Song*. Richard K. Walton and Robert W. Lawson, *Backyard Bird Song: A Guide to Bird-song Identification*, compact disc (New York: Houghton Mifflin, 1991), 2. And see the attributions of many birds on, for example, Allaboutbirds.org, including the wood thrush, bobolink, scarlet tanager, wrentit, Baltimore oriole, golden plover, and many others (https://www.allaboutbirds.org, accessed Sept. 12, 2014).

20. Bruynincxk, "Sounds Sterile."

21. J. Harold. Ennis, "American Bird Songs by Laboratory of Ornithology, Cornell University," *Bios* 13, no. 4 (1942): 268; Bentley Glass, "American Bird Songs," *Quarterly Review of Biology* 20, no. 3 (1945): 279.

22. Glass, "American Bird Songs," 279.

23. Another example can be found in the different representations of the tree pipit in Michael Lohmann and Jean C. Roché, *Guide des Oiseaux Chanteurs* (Aartselaar: Chantecler, 2005). Here the bird is first introduced by a short strophe, reproduced in a spec-

trogram on page 10, meant to represent the species; but in the accompanying CD, the much longer recording gives a different and far more complex impression of the species.

24. Les Beletsky, *Bird Songs: 250 North American Birds in Song* (San Francisco: Chronicle, 2006), 10, 107.

25. Lohmann and Roché, *Guide des Oiseaux Chanteurs*, 36.

26. Beletsky, *Bird Songs*, 250.

27. Arthur A. Allen, *The Golden Plover and Other Birds* (Ithaca, NY: Comstock, 1939), 5.

28. Peterson, *A Field Guide to the Birds of Eastern and Central North America*, 9.

29. The literal geography of the garden was first suggested to me by an unpublished manuscript by Professor Jon K. Barlow, which detailed the parallels between medieval paradisical cartography and Mount Kilimanjaro.

30. Jean Delumeau, *History of Paradise: The Garden of Eden in Myth and Tradition*, trans. Matthew O'Connell (New York: Continuum, 2000), 39–44.

31. For historical descriptions of the garden, see Delumeau, *History of Paradise*; Alessandro Scafi, *Mapping Paradise: A History of Heaven on Earth* (Chicago: University of Chicago Press, 2006); Alessandro Scafi, *Maps of Paradise* (Chicago: University of Chicago Press, 2014).

32. Delumeau, *History of Paradise*, 44.

33. Scafi, *Mapping Paradise*, 286.

34. ibid.

35. The story, a midrash later appropriated as a Christian narrative, is told in Wilfred Schuchat, *The Garden of Eden and the Struggle to be Human: According to the Midrash Rabbah* (New York: Devora, 2006), 345.

36. C. S. Lewis, *The Magician's Nephew* (New York: Harper Collins, 1994), 190.

37. Ernest Hemingway, *The Garden of Eden* (New York: Scribner, 1987), 3.

38. Octavia Butler, *Lilith's Brood* (New York: Warner Books, 1989), 345.

39. Steven Feld, *Rainforest Soundwalks*, liner notes (EarthEar ee1062, 2001), 3.

40. Miyoko Chu, *Birdscapes: A Pop-up Celebration of Bird Songs in Stereo Sound* (San Francisco: Chronicle, 2008), Sonoran Desert.

41. Yi-Fu Tuan, "Sign and Metaphor," *Annals of the Association of American Geographers* 68, no. 3 (Sept. 1978), 372.

42. He goes on in the same passage to say that "we know nothing of the mental conditions in which musical creation takes place . . . the musical creator is a being comparable to the gods, and music itself the supreme mystery of the science of man, a mystery that all the various disciplines come up against and which holds the key to their progress." Claude Lévi-Strauss, *The Raw and the Cooked: Mythologiques*, vol. 1, trans. John and Doreen Weightman (Chicago: University of Chicago Press, 1983), 17, 18.

43. Steven Feld, *Sound and Sentiment: Birds, Weeping, Poetics, and Song in Kaluli Expression* (Philadelphia: University of Pennsylvania Press), 1982. The book was first

submitted as a doctoral dissertation to Indiana University in 1979; the first American edition of Strauss's book was in 1969, ten years earlier. Claude Lévi-Strauss, *The Raw and the Cooked* (New York: Harper and Row, 1969).

44. Lévi-Strauss, *The Raw and the Cooked* (1983), 19; see note 7.

45. Ibid., 18.

46. Robert Sherlaw-Johnson, *Messiaen* (Berkeley: University of California Press, [1975] 1989); see also William Austin's mixed review, which may reflect a broader sense that Messiaen's advocates were not giving substantive responses to the critics of his bird style. "*Messiaen* by Robert Sherlaw-Johnson," *Notes* 33, no. 1 (Sept. 1976): 61–62.

47. His student Alexander Goehr recalls giggling with other students at the premiere of *Oiseaux exotiques*, to Messiaen's frustration. Recounted in Peter Hill and Nigel Simeone, *Messiaen* (New Haven: Yale University Press, 2005), 217.

48. Claude Samuel, *Entretiens avec Olivier Messiaen* (Paris: Belfond, 1967), 65.

49. Trevor Hold, "Messiaen's Birds," *Music and Letters* 52, no. 2 (Apr. 1971): 113–14.

50. Peter Hill, "From Réveil des oiseaux to Catalogue d'oiseaux: Messiaen's Cahiers de notations des chants d'oiseaux, 1952–59," in *Messiaen Perspectives I: Sources and Influences*, ed. Christopher Dingle and Robert Fallon (Burlington, VT: Ashgate, 2013), 145. See also Hold, "Messiaen's Birds"; Sherlaw-Johnson, *Messiaen*; Jacques Penot, "Olivier Messiaen ornithologue," in *Portrait(s) d'Olivier Messiaen*, ed. Catherine Massip (Paris: Bibliothèque Nationale de France, 1996), 71; Robert Sherlaw Johnson, "Birdsong," in *The Messiaen Companion*, ed. Peter Hill (Portland, OR: Amadeus, 1995), 249–65; Peter Hill and Nigel Simeone, *Oiseaux Exotiques* (Burlington, VT: Ashgate), 2007; Robert Fallon, "The Record of Realism in Messiaen's Bird Style," in *Olivier Messiaen: Music, Art, and Literature*, ed. Christopher Dingle and Nigel Simeone (Burlington, VT: Ashgate, 2007), 115–36.

51. Annie Petit, "L'esprit de la science anglaise et les Français au XIXème siècle," *British Journal for the History of Science* 17, no. 3 (Nov. 1984): 273–93.

52. Chansigaud, *Histoire de l'ornithologie*, 203–6.

53. The *ornithologue* is the "plus savants, très reliés à la science ornithologique; ornithologiste à tous les passionnés dont je suis et qui s'efforcent de mieux faire connaître les oiseaux pour les mieux protéger." Penot, "Olivier Messiaen ornithologue," 71–72.

54. Sherlaw-Johnson, *Messiaen*, 116; see also Samuel, *Entretiens avec Olivier Messiaen*, 91.

55. "Enfin, si le laboratoire permet au travailler d'approfondir certaines problèmes, ou d'apporter à certaines questions des éléments d'une précision mathématique, on sait que la nature se plait à derouter les constructions de l'esprit les plus objectifs, il ne faut donc pas que l'amour du microscope ou le culte des expériences dirigées, des statistiques et des courbes graphiques fassent perdre le contact du réel. C'est ce contact que l'observateur in natura est chargé de maintenir et ce n'est pas sa tâche la plus aisée bien au contraire." R.-D. Etchecopar, "Reflexions sur l'utilité de l'observation 'in natura' même en systematique," *L'Oiseau et la Revue Française d'Ornithologie* 21, no. 4 (1951): 311).

56. Robert Fallon, "Messiaen's Mimesis: The Language and Culture of the Bird Styles" (PhD diss., University of California–Berkeley, 2005), 30–31.

57. Jacques Delamain, *Pourquoi les oiseaux chantent* (Paris: Stock, 1948), 22, 26.

58. "Sous le soleil de midi, la phrase monotone et traînante de l'Ortolan a résonné seule, dans les vignobles pleins de lumière. Avec la brise du soir, les sons ont repris. Puis, au déclin du jour, le bavardage confus des petites voix sans art s'est éteint. Le Rossignol a chanté encore, par fragments de strophes, sans conviction. Alors, le Loriot a sifflé une dernière fois. Des voix après la sienne, sont montées de la paix du soir, discrètes, rares, comme imprégnées de silence et de nuit. Un Merle a tenu la scène." Jacques Delamain, *Porqoui les Oiseaux Chantent* (Paris: Stock, 1964), 30.

59. See, for example, Jacques Delamain, *Portraits d'Oiseaux*, vol. 2, illustrated by Roger Reboussin (Paris: Stock, 1952), 78; Olivier Messiaen, "L'Alouette Lulu," in *Catalogue d'oiseaux*, vol. 3 (Paris: Leduc, 1964), 1. There are several cases of this, some of which have been detailed by Christopher Brent Murray in "Les metaphores vegetales dans le *Catalogue d'oiseaux* d'Olivier Messiaen," unpublished manuscript.

60. Olivier Messiaen, "Le Chocard des Alpes," in *Catalogue d'oiseaux*, book 1, no. 1 (Paris: Leduc, 1964), 1, 2, 11, 12.

61. Wai-Ling Cheong has pointed out that such parallel motion is usually a clue to colored harmonies in Messiaen's work. "Messiaen's Triadic Coloration: Modes as Intervention," *Music Analysis* 21 (2002): 53–84. The recent publication of the seventh volume of Messiaen's treatise (the one on color) also makes many of us lick our chops when looking at examples that, like the sunrise in "Le Loriot," do not easily reduce to modal analysis.

62. Olivier Messiaen, "Le Loriot," in *Catalogue d'oiseaux*, book 1, no. 2 (Paris: Leduc, 1964), 1, 4, 5, 9–10.

63. Olivier Messiaen, "L'Alouette calandrelle" and "La bouscarle," in *Catalogue d'oiseaux*, book 5, no. 8 (Paris: Leduc, 1964).

64. Fallon, "The Record of Realism in Messiaen's Bird Style."

65. http://www.earthear.com/catalog/rainsoundwalks.html, accessed May 2008.

66. Review by Carlos Palombini in *Leonardo Digital Reviews*, June 2001, accessed May 2008, http://www.leonardo.info/reviews/jun2001/cd_SOUNDWALKS_palombini.html.

67. Feld, *Rainforest Soundwalks*, liner notes, 5.

68. Schafer, *Soundscape*, 139–44.

69. Feld, *Rainforest Soundwalks*, liner notes, 3.

70. Ibid.

71. A "seyak" frame of reference is one of many possibilities even among songbirds. In Manhattan the robins are performing such extended and varied songs every morning between 3:00 and 5:00 a.m. in the spring.

72. Accessed Mar. 3, 2010, Mar. 4, 2010, and June 15, 2010.

73. Roy Thomas, *Stan Lee's Amazing Marvel Universe* (New York: Sterling, 2006). My

information about Becker and Mayer's development of the audio series comes from a telephone conversation with Peter Schumacher, head of Becker and Mayer's adult product division, Feb. 27, 2010.

74. Beletsky, *Bird Songs*.

75. These include Donald Kroodsma's *The Backyard Birdsong Guide: Eastern and Central North America* (San Francisco: Chronicle, 2008), and his *The Backyard Birdsong Guide: Western North America* (San Francisco: Chronicle, 2008).

76. I corresponded briefly with both Beletsky and Kroodsma, and they confirmed for me that Beletsky primarily wrote text and left the sound to his colleagues at Cornell, while Kroodsma oversaw all the details of the sound elements in his book.

77. Kroodsma, *The Backyard Birdsong Guide*, 48.

78. Chu, *Birdscapes*.

79. An intriguing point of comparison is Hollis Taylor's elaborate reconstruction in text of the habitat of the Australian butcherbird in *Is Birdsong Music? Outback Encounters with an Australian Songbird*, a remarkable sonic ethnography that can be compared in its own right to both Feld's and Chu's projects.

80. Telephone conversation with Miyoko Chu, Dec. 19, 2013.

81. A recent example is the National Audubon Society's nationwide release in September 2014 of a study on climate change that used publicly reported bird counts to calculate shifting migration patterns affected by climate, claiming that nearly half of North American bird species are currently threatened by climate change (http://climate.audubon.org, accessed Sept. 15, 2014).

82. Most counts allow birders to count any bird "conclusively identified by sight or sound." See the rules for the American Birding Association's "Big Day Count," accessed Sept. 18, 2014, http://listing.aba.org/big-day-count-rules/. See also National Audubon Society, *Audubon Christmas Bird Count Compiler's Manual* (Willow Grove, PA: Audubon Science, 2014), 4.


CONCLUSION  *The Animanities*

1. "The Song Sparrow sings a loud, clanking song of 2–6 phrases that typically starts with abrupt, well-spaced notes and finishes with a buzz or trill." https://www.allaboutbirds .org/guide/Song_Sparrow/sounds, accessed Aug. 29, 2016.

2. Una Chaudhuri, "(De)Facing the Animals: Zooësis and Performance," *Drama Review* 51, no. 1 (Spring 2007): 8.

3. These are references, respectively, to Cary Wolfe, ed., *Zoontologies: The Question of the Animal* (Minneapolis: University of Minnesota Press, 2003); Jennifer Wolch, "Zoopolis," in *Animal Geographies: Place, Politics, and Identity in the Nature-Culture Borderlands*, ed. Jennifer Wolch and Jody Emel (New York: Verso, 1998), 119–38; J. M. Olson and Kathleen

Hulser, "Petropolis: A Social History of Urban Animal Companions," *Visual Studies* 18, no. 2 (2003): 133–43; Jacques Derrida, "'Eating Well' or the Calculation of the Subject," in *Who Comes After the Subject?*, ed. Eduardo Cadava, Peter Connor, and Jean-Luc Nancy (New York: Routledge, 1991), 96–119; and groups such as http://www.learning-animals .org/en/organization (accessed Nov. 22, 2015).

4. "Zoomusicology" was coined by Messiaen's student François-Bernard Mâche in *Music, Myth, and Nature or, The Dolphins of Arion*, trans. Susan Delaney (Philadelphia: Harwood Academic, 1992). "Ornithomusicology" is attributed to Hungarian song collector and recordist Peter Szöke. "Ornitomuzikológia," *Magyar Tudomany* 9 (1963): 592–607. More recently, "ecomusicology" has been used by a number of American music scholars, notably Aaron Allen. Aaron S. Allen, ed., "Colloquy: Ecomusicology: Ecocriticism and Musicology," *Journal of the American Musicological Society* 64, no. 2 (Summer 2011): 391–94.

5. Chaudhuri, "(De)Facing the Animals," 9.

6. This shift is made evident in Jane Bennett's discourse of vibrant materiality, in Rosi Braidotti's endorsement of *zoe*, in Elizabeth Povinelli's geontopower, and more. See Jane Bennett, *Vibrant Matter: A Political Ecology of Things* (Durham, NC: Duke University Press, 2010); Rosi Braidotti, *The Posthuman* (Malden, MA: Polity, 2013), Elizabeth Povinelli, *Geontologies: A Requiem to Late Liberalism* (Durham, NC: Duke University Press, 2016), and Richard Grusin, ed., *Anthropocene Feminism* (Minneapolis: University of Minnesota Press, 2017).

7. Although they do not adopt (or know about) my imaginary field, the project of locating issues of rights and dignity at the center of studies of the more-than-human world has many allies, including recent examples such as Mel Y. Chen, *Animacies: Biopolitics, Racial Mattering, and Queer Affect* (Durham, NC: Duke University Press, 2012); Timothy Morton, *Hyperobjects: Philosophy and Ecology After the End of the World* (Minneapolis: University of Minnesota Press, 2013); François Laruelle, *General Theory of Victims*, trans. Jessie Hock and Alex Dubilet (Malden, MA: Polity Press, 2015).

8. Judith Butler, *Senses of the Subject* (New York: Fordham University Press, 2015), 6.

9. See, for example, Philip Auslander's study of legal cases involving individual and corporate voices in *Liveness: Performance in a Mediatized Culture* (New York: Routledge, 1999).

10. Such cases include Cetacean Community v. George W. Bush (2003), which argued that the US Navy's use of low-frequency sonar violated existing conservation laws; Tilikum et al. v. Sea World Parks and Entertainment Inc. et al. (2012), a case sponsored by PETA asking for recognition that the confinement of five wild-captured Sea World orcas constituted a violation of the Thirteenth Amendment's prohibition of slavery; People ex. rel. Nonhuman Rights Project, Inc. v. Lavery (2014), in which Steven Wise's Nonhuman Rights Project applied for a writ of habeus corpus on behalf of a captive chimpanzee

named Tommy in Fulton County, New York, attempting to establish Tommy's legal personhood; and a similar case on behalf of two research chimps at Stony Brook University, Hercules and Leo, in 2015. Matter of Nonhuman Rights Project Inc. v Stanley (2015). So far largely unsuccessful, these cases have attempted to establish legal personhood for nonhuman animals that legal defenders can compare with human rationality. In the past, these cases have been complicated by attempts to compare the challenges of establishing legal guardianship for cognitively impaired human adults with the legal representation of nonhuman animals. See also http://www.nonhumanrights.org, Steven Wise's website for the Nonhuman Rights Project.

11. Dinesh Wadiwel discusses aspects of this dual view of personhood in constructions of slavery and food production in his *The War against Animals* (Boston: Brill, 2015).

12. The interplay of personhood, guilt, and innocence is explored in Karen M. Morin's "Carceral Space: Prisoners and Animals," *Antipode* 48, no. 5 (Nov. 2016): 1317–36; and Lori Gruen, ed., *The Ethics of Captivity* (New York: Oxford University Press, 2014).

13. Joan Wallach Scott, *Gender and the Politics of History* (New York: Columbia University Press, 1988).

14. José Esteban Muñoz, "Theorizing Queer Inhumanisms: The Sense of Brownness," *GLQ* 21, no. 2–3 (June 2015): 209–10.

15. The remarkable scope and relevance of notions of difference and identity is indicated by, among other things, the enormous range covered by, for example, frameworks drawn from notions of grammar, species, philosophy, subjectivity, queer utopias; and victimhood. See Jacques Derrida, "Differánce," trans. Alan Bass, *Margins of Philosophy* (Chicago: University of Chicago Press, 1982), 3–27; Cary Wolfe, ed., *Zoontologies: The Question of the Animal* (Minneapolis: University of Minnesota Press, 2003); Etienne Balibar, *Identity and Difference: John Locke and the Invention of Consciousness*, trans. Warren Montag (New York: Verso, 2013); Butler, *Senses of the Subject*; José Estevan Muñoz, *Cruising Utopia: The Then and There of Queer Futurity* (New York: New York University Press, 2009); Laruelle, *General Theory of Victims*.

16. Una Chaudhuri, "Introduction," in Chaudhuri and Holly Hughes, eds., *Animal Acts: Performing Species Today* (Ann Arbor: University of Michigan Press, 2014), 8.

17. A useful overview of posthumanism is provided in the introduction to Cary Wolfe, *What Is Posthumanism?* (Minneapolis: University of Minnesota Press, 2010).

18. See, for example, Michel Foucault, *The Order of Things: An Archaeology of the Human Sciences*, trans. Alan Sheridan (New York: Vintage, 1994); Bruno Latour, *We Have Never Been Modern*, trans. Catherine Porter (Cambridge, MA: Harvard University Press, 1993). See also Étienne Balibar, "Citizen Subject," in *Who Comes After the Subject?*, ed. Eduardo Cadava, Peter Connor, and Jean-Luc Nancy (New York: Routledge, 1991), 33–57.

19. These scholars include Kalpana Rahita Seshadri, *HumAnimal: Race, Law, Language* (Minneapolis: University of Minnesota Press, 2012); and Alexander Weheliye, *Habeus*

*Viscus: Racializing Assemblages, Biopolitics, and Black Feminist Theories of the Human* (Durham, NC: Duke University Press, 2014).

20. See Paul J. Crutzen and Eugene F. Stoermer, "The Anthropocene," *Global Change Newsletter* 41 (2000): 17–18; and Grusin, *Anthropocene Feminism*.

21. See Donna Haraway, *Primate Visions: Gender, Race, and Nature in the World of Modern Science* (New York: Routledge, 1989); and *Aph Ko and Syl Ko, Aphro-isms: Essays on Pop Culture, Feminism, and Black Veganism from Two Sisters* (New York: Lantern), 2017.

22. See, for example, Nathaniel Comfort, *The Science of Human Perfection: How Genes Became the Heart of American Medicine* (New Haven: Yale University Press, 2012), 240–43. See also Jenny Reardon, *Race to the Finish: Identity and Governance in an Age of Genomics* (Princeton, NJ: Princeton University Press, 2005).

23. Puping Liang et al., "CRISPR/Cas9-mediated gene editing in human tripronuclear zygotes," *Protein and Cell* 6, no. 5 (May 2015): 363–72. The article argues that there is a pressing need to develop further research to prevent problematic gene editing, but its authors do not appear familiar with the history of either Western or Asian eugenics.

24. See, for example, https://www.pandora.com/about/mgp, accessed Nov. 27, 2015.

# BIBLIOGRAPHY

## Archival Sources

Charles C. Adams Papers. Wallace Craig Correspondence. Western Michigan University Archives and Regional History Collections, Kalamazoo, MI.

Charles Davenport Papers. Mss. B. D27. American Philosophical Society Library, Philadelphia, PA.

Clarence Weed Correspondence. MS W444. Special Collections, American Museum of Natural History, New York, NY.

Correspondence Regarding the Wood Pewee. Box A 3157–55. New York State Archives, Albany, NY.

E. M. von Hornbostel Papers. Max Wertheimer Papers. Manuscripts and Archives Division, New York Public Library. Astor, Lenxo, and Tilden Foundations, New York, NY.

Eugenics Record Office Papers. American Philosophical Society Library, Philadelphia, PA.

Laura Boulton Collection. Archives for Traditional Music, University of Indiana, Bloomington, IN.

Leonard Carmichael Papers. Mss B. C212. American Philosophical Society, Philadelphia, PA.

1931 Pulitzer Angola Expedition Papers. 2006–8. Carnegie Museum of Natural History Archives, Pittsburgh, PA.

Robert Mearns Yerkes Papers. Sterling Memorial Library Manuscripts and Archives, Yale University, New Haven, CT.

Straus Correspondence. American Museum of Natural History Central Archives, New York, NY.

## Primary Printed Sources

Abraham, Otto, and Erich M. von Hornbostel. "Suggested Methods for the Transcription of Exotic Music." Translated by George List and Eve List. *Ethnomusicology* 38, no. 3 (Autumn 1994): 425–56. Originally published as "Vorschläge zur Transkription exotischer Melodien," *Sammelbände der Internationalen Musikgesellschaft* 2, no. 1 (1909–10): 1–25.

Adams, Charles C. "Introductory Note: Bird Song and the State Museum." In *The Song of the Wood Pewee Myochanes virens linnaeus: A Study of Bird Music*, 7–8. New York State Museum Bulletin 334, June 1943.

Adler, Guido. "Guido Adler's 'The Scope, Method, and Aim of Musicology,' (1885): An English Translation with an Historico-Analytical Commentary." Translation and commentary by Erica Mugglestone. *Yearbook for Traditional Music* 13 (1981): 1–21. Originally published as "Umfang, Methode und Ziel der Musikwissenschaft," *Vierteljahrsschrift für Musikwissenschaft* 1 (1885): 5–20.

———. "Internationalism in Music." Translated by Theodore Baker. *Musical Quarterly* 11 (1925): 281–300.

———. *Der Stil in der Musik*. Vol. 1. Leipzig: Breitkopf and Härtel, 1911.

Adorno, Theodor. *The Culture Industry: Selected Essays on Mass Culture*. Edited by J. M. Bernstein. New York: Routledge, 1991.

Agassiz, Alexander. *Letters and Recollections of Alexander Agassiz*. Edited by George Russell Agassiz. New York: Houghton Mifflin, 1913.

Alland, Alexander. *Evolution and Human Behavior*. New York: Natural History Press, 1967.

Allard, Harry A. *Our Insect Instrumentalists and Their Musical Technique*. Washington, DC: United States Government Printing Office, from the Smithsonian Report for 1928, 1929.

Allen, Arthur A. *The Golden Plover and Other Birds*. Ithaca, NY: Comstock, 1939.

Allen, Arthur Augustus, and P. P. Kellogg. *American Bird Songs*. Ithaca, NY: Comstock, 1942. 78 rpm.

Allen, Francis H. "Review: The Song of the Wood Pewee." *Wilson Bulletin* 56, no. 1 (Mar. 1944): 52–54.

———. "Group Variation and Bird-Song." *Auk* 40, no. 4 (Oct. 1923): 643–49.

———. "The Evolution of Bird-Song." *Auk* 36, no. 4 (Oct. 1919): 528–36.

———. "The Aesthetic Sense in Birds as Illustrated by the Crow." *Auk* 36, no. 1 (Jan. 1919): 112–13.

Allen, Grant. "Aesthetic Evolution in Man." *Mind* 5, no. 20 (Oct. 1880): 445–64.

Allen, Warren Dwight. "Review: A History of Western Music by Donald Jay Grout." *Journal of Research in Music Education* 8, no. 2 (Autumn 1960): 124–26.

———. *Philosophies of Music History*. New York: American Book, 1939.

Anonymous. "British Folk-Song." *Musical Times* 45, no. 733 (Mar. 1904): 175–76.

Anonymous. "The Collecting of English Folk-Songs." *Musical Times* 46, no. 746 (Apr. 1905): 258.

Austin, William. "*Messiaen* by Robert Sherlaw-Johnson." *Notes* 33, no. 1 (Sept. 1976): 61–62.

———. "The Idea of Evolution in the Music of the Twentieth Century." *Musical Quarterly*

39, no. 1 (Jan. 1953): 26–36.

Baker, Theodore. *Biographical Dictionary of Musicians.* New York: Schirmer, 1905.

Baptista, Luis F., and Sandra L. Gaunt. "Advances in Studies of Avian Sound Communication." *Condor* 96, no. 3 (Aug. 1994): 817–30.

Barry, Phillips. "Folk-Music in America." *Journal of American Folklore* 22, no. 83 (1909): 72–81.

Bartók, Béla. *Béla Bartók, Essays.* Edited by Benjamin Suchoff. Lincoln, NE: Bison, 1992.

――――. *Serbo-Croatian Folk Songs.* Edited and translated by Albert B. Lord. New York: Columbia University Studies in Musicology, 1951.

Batchelder, Frederick. "The Transcription of Bird Songs." *Primary Education* 12, no. 3 (May 1904): 216–17.

Beaglehole, Ernest, Juan Comas, L. A. Costa Pinto, Franklin Frazier, Morris Ginsberg, Humayun Kabir, Claude Levi-Strauss, and Ashley Montagu. "Statement on Race: 1950." In *Four Statements on the Race Question,* 30–35. Paris: United Nations, 1969.

Bechert, Paul. "The Donaueschingen Festival." *Musical Times* 66, no. 991 (Sept. 1925): 844–45.

Beletsky, Les. *Bird Songs: 250 North American Birds in Song.* San Francisco: Chronicle, 2006.

Bely, Andrei. *Gogol's Artistry.* Translated by Christopher Colbath. Evanston, IL: Northwestern University Press, 2009.

Benedict, Ralph Cürtiss. "The Type and Identity of Dryopteris Clintoniana (D. C. Eaton) Dowell." *Torreya* 9, no. 7 (July 1909): 133–40.

Bergson, Henri. *Creative Evolution.* Translated by Arthur Mitchell. New York: Holt, 1913.

Bigelow, Anna Nieglieh, and Maurice Alpheus Bigelow. *Applied Biology: An Elementary Textbook and Laboratory Guide.* New York: Macmillan, 1917.

Blakeslee, Albert. "Ideals of Science." *Science* 92, no. 2400 (Dec. 27, 1940): 589–92.

――――. "Corn and Education." *Journal of Education* 88, no. 12 (Oct. 3, 1918): 320–21, 326.

Boas, Franz. "Teton Sioux Music." *Journal of American Folklore* 38, no. 148 (1925): 319–24.

Boulton, Laura. *The Music Hunter.* New York: Doubleday, 1969.

Boulton Craytor, Laura T., and Charles Davenport. "Comparative Social Traits of Various Races." *Journal of Applied Society* 7 (1923): 127–34.

Bourdieu, Pierre. "The Aristocracy of Culture." Translated by Richard Nice. *Media Culture Society* 2 (1980): 225–54.

Boyd, William C. "The Use of Genetically Determined Characters, Especially Serological Factors Such as Rh, in Physical Anthropology." *Southwestern Journal of Anthropology* 3, no. 1 (Spring 1947): 32–49.

Brand, Albert R. "Why Bird Song Can Not Be Described Adequately." *Wilson Bulletin* 49, no. 1 (Mar. 1937): 11–14.

———. "The 1935 Cornell-American Museum Ornithological Expedition." *Scientific Monthly* 41, no. 2 (Aug. 1935): 187–90.

———. "A Method for the Intensive Study of Bird Song." *Auk* 52, no. 1 (Jan. 1935): 40–52.

———. *Songs of Wild Birds*. New York: Thomas Nelson, 1934.

———. "Recording Sounds of Wild Birds." *Auk* 49, no. 4 (1932): 436–39.

Bray, Bruce. "Review: A History of Western Music by Donald Jay Grout." *Music Educators Journal* 46, no. 6 (June–July 1960): 78.

Broadwood, Lucy, and Cecil J. Sharp. "Some Characteristics of English Folk-Music." *Folklore* 19, no. 2 (June 1908): 132–52.

Broughton, Simon, and Mark Ellingham, eds. *The Rough Guide to World Music: Africa and Middle East*. London: World Music Network, 2007.

Brown, Mary Elizabeth. *Preliminary Catalogue of the Crosby-Brown Collection*. New York: Metropolitan Museum of Art, 1901.

Butler, Octavia. *Lilith's Brood*. New York: Warner, 1989.

Campbell, Joseph. *The Hero with a Thousand Faces*. 3rd ed. Novato, CA: New World Library, 2008.

Carlyle, Thomas. *On Heroes, Hero-Worship, and the Heroic in History*. London: James Fraser, 1841.

Chamberlain, Houston Stewart. *Richard Wagner*. London: Dent, 1900.

Chansigaud, Valérie. *Histoire de l'ornithologie*. Paris: Delachaux et Niestlé, 2007.

Cheney, Simeon Pease. *Wood Notes Wild: Notations of Bird Music*. Boston: Lee and Shepard, 1891.

Chilton, Carroll Brent. *The De-assification of Music: A Propagandist Magazine of One Number, Containing News of Importance to All Music Lovers, Especially to All Owners of Player Pianos*. New York: Chilton, 1922.

Chu, Miyoko. *Birdscapes: A Pop-up Celebration of Bird Songs in Stereo Sound*. San Francisco: Chronicle, 2008.

Clark, Xenos. "Animal Music, Its Nature and Origin." *American Naturalist* 13, no. 4 (Apr. 1879): 209–23.

Coffin, Lucy V. Baxter. "Individuality in Bird Song." *Wilson Bulletin* (June 1928): 95–99.

Combarieu, Jules. *La Musique, Ses Lois, son Évolution*. Paris: Flammarion, 1907.

Cooley, Timothy, Davide Locke, David P. McAllester, Anne K. Ramussen, and Jeff Todd Titon. *Worlds of Music: An Introduction to the Music of the World's Peoples*. 5th ed. Cengage Learning, 2008. eBook.

Cornish, Charles John. *Life at the Zoo: Notes and Traditions of the Regent's Park Gardens*. London: Seeley, 1896.

Coues, Professor, and S. S. Oregon. "On Some New Terms Recommended for Use in Zoological Nomenclature." *Auk* 1, no. 4 (Oct. 1884): 320–22.

Craig, Wallace. *The Song of the Wood Pewee Myochanes Virens Linnaeus: A Study of Bird*

*Music.* Albany: New York State Museum Bulletin no. 334, June 1943.

———. "Request for Data on the Twilight Song of the Wood Pewee." *Science* 63, no. 1638 (May 1926): 525.

———. "The Twilight Song of the Wood Pewee: A Preliminary Statement." *Auk* 43, no. 2 (Apr. 1926): 150–52.

———. "A Note on Darwin's Work on the Expression of the Emotions in Man and Animals." *Journal of Psychology and Social Psychology* 16, no. 5 (Apr. 1921): 356–66.

———. "The Expression of Emotion in the Pigeons. III. The Passenger Pigeon (Ectopistes migratorius Linn.)." *Auk* 28, no. 4 (Oct. 1911): 408–27.

———. "The Expression of Emotion in the Pigeons. II. The Mourning Dove (Zenaidura macruora Linn.)." *Auk* 28, no. 4 (Oct. 1911): 398–407.

———. "The Expression of Emotion in the Pigeons. I. The Blond Ring-Dove." *Journal of Comparative Neurology and Psychology* 19, no. 1 (Apr. 1909): 29–82.

———. "The Voices of Pigeons Regarded as a Means of Social Control." *American Journal of Sociology* 14, no. 1 (July 1908): 86–100.

Crocker, Richard. *A History of Musical Style.* New York: Dover, [1966] 1986.

Crow, James F. "Birth Defects, Jimson Weeds and Bell Curves." *Genetics* 147 (Sept. 1997): 1–6.

Cuvier, Georges. *The Animal Kingdom, Arranged According to Its Organization.* Translated by H. McMurtrie. London: Orr and Smith, 1834.

Darwin, Charles. *The Descent of Man, and Selection in Relation to Sex.* New York: Barnes and Noble, [1871] 2004.

———. *The Expression of the Emotions in Man and Animals.* London: John Murray, 1872.

Daubeny, Ulric. "Gramophone 'Why Nots?'" *Musical Times* 61, no. 929 (July 1920): 486–87.

Davenport, Charles Benedict. "Comparative Social Traits of Various Races." *School and Society* 14, no. 340 (1921): 344–48.

———. "The Personality, Heredity, and Work of Charles Otis Whitman, 1843–1910." *American Naturalist* 51, no. 601 (Jan. 1917): 5–30.

———. *The Science of Human Improvement by Better Breeding.* New York: Holt, 1910.

Davis, L. Irby. "Biological Acoustics and the Use of the Sound Spectrograph." *Southwestern Naturalist* 9, no. 3 (Aug. 20, 1964): 119–20.

Delamain, Jacques. *Portraits d'Oiseaux.* Vol. 2. Illustrated by Roger Reboussin. Paris: Stock, 1952.

———. *Pourquoi les oiseaux chantent.* Paris: Stock, 1948.

*The Demonstration Collection of E. M. von Hornbostel and the Berlin Phonogramm-Archiv.* New York: Ethnic Folkways, 1963. 78 rpm.

Densmore, Frances. *Teton Sioux Music.* Washington, DC: Smithsonian Institution Bureau of American Ethnology Bulletin 61, 1918.

———. *Chippewa Music II*. Washington, DC: Smithsonian Institution Bureau of American Ethnology Bulletin 53, 1913.

Dent, Edward J. "The Romantic Spirit in Music." *Proceedings of the Musical Association*, 59th Sess. (1932–33): 85–102.

Dickinson, Edward. *The Study of the History of Music*. New York: Scribner, 1920.

D'Indy, Vincent, and Auguste Sérieyx. *Cours de Composition Musicale*. Vol. 1. Paris: Durand, 1912.

Dobzhansky, Theodosius. "Race and Humanity." *Science*, new series 113, no. 2932 (Mar. 9, 1951): 264–66.

Doyle, Arthur Conan. "A Study in Scarlet." In *Sherlock Holmes: The Complete Novels and Stories*, vol. 1. Introduction by Loren Estleman. New York: Bantam, [1887] 1986.

Dunn, James O. "The Passenger Pigeon in the Upper Mississippi Valley." *Auk* 12, no. 4 (Dec. 1895): 389.

Eaton, Elon Howard, and Louis Agassiz Fuertes. *Birds of New York*. New York: New York State Museum, 1910.

Elliott, Lang, Donald Stokes, and Lilian Stokes. *Stokes Field Guide to Birdsongs*. Time Warner Audio Books, 2-32482, 1997. Compact disc set.

Ellis, Alexander J. "On the Musical Scales of Various Nations." *Journal of the Society for the Arts* 33, no. 1688 (Mar. 27, 1885): 492–500.

Ennis, J. Harold. "American Bird Songs by Laboratory of Ornithology, Cornell University." *Bios* 13, no. 4 (1942): 268.

Entwistle, William J. "Notation for Ballad Melodies." *Publications of the Modern Language Association* 55, no. 1 (Mar. 1940): 61–72.

Eschman, Karl H. "The Rhetoric of Modern Music." *Musical Quarterly* 7, no. 2 (Apr. 1921): 157–66.

Etchecopar, R.-D. "Reflexions sur l'utilité de l'observation 'in natura' même en systematique." *L'Oiseau et la Revue Française d'Ornithologie* 21, no. 4 (1951): 310–16.

"Fallacies of Racism Exposed." *Unesco Courier* 3, no. 6 (July 1950): 1.

Farnsworth, Paul R. "Sacred Cows in the Psychology of Music." *Journal of Aesthetics and Art Criticism* 7, no. 1 (Sept. 1948): 48–51.

Feld, Steven. *Rainforest Soundwalks*. EarthEar ee1062, 2001. Compact disc.

———. *Sound and Sentiment: Birds, Weeping, Poetics, and Song in Kaluli Expression*. Philadelphia: University of Pennsylvania Press, 1982.

de Fénis, François. *Les Langages des Animaux et de l'homme*. Hanoi: Extrême-orient, 1921.

Fernald, James Champlin. *The Concise Standard Dictionary of the English Language*. New York: Funk and Wagnalls, 1910.

Fernald, M. L. "The Identity of Angelica Lucida." *Rhodora* 21, no. 248 (Aug. 1919): 144–47.

Ferris, G. F. *The Principles of Systematic Entomology*. Stanford University Biological Sciences, series 5, no. 3. San Francisco: Stanford University Press, 1928.

Fewkes, Jesse Walter. "On the Use of the Phonograph in the Study of the Languages of American Indians." *Science* 15, no. 378 (May 2, 1890): 267–69.

Fisher, William Arms. "What is Music?" *Musical Quarterly* 15, no. 3 (July 1929): 360–70.

Fowler, H. W., and F. G. Fowler, eds. *The Concise Oxford Dictionary of Current English.* New York: Oxford University Press, 1911.

Galpin, F. W. "The Whistles and Reed Instruments of the American Indians of the North-West Coast." *Proceedings of the Musical Association* 29 (1902): 115–38.

Gardiner, William. *The Music of Nature.* Boston: Wilkins and Carter, 1841.

Garstang, Walter. *Songs of the Birds.* London: John Lane, 1922.

Gillet, W. *The Phonograph, and How to Construct It.* New York: Spon and Chamberlain, 1892.

Gilliland, Esther Goetz. "The Healing Power of Music." *Music Educators Journal* 31, no. 1 (Sept.–Oct. 1944): 18–20.

Gilman, Benjamin Ives. "The Science of Exotic Music." *Science* 30, no. 772 (Oct. 15, 1909): 532–35.

———. *Hopi Songs.* Cambridge, MA: Riverside, 1908.

Gilman, Lawrence. "Music of the Month: A Tone Poet from Italy." *North American Review* 215, no. 795 (Feb. 1922): 267–72.

Ginsberg, Morris. "Social Evolution." In *Darwinism and the Study of Society,* ed. Michael Banton, 101–2. Chicago: Quadrangle, 1961.

Glass, Bentley. "American Bird Songs." *Quarterly Review of Biology* 20, no. 3 (1945): 279.

Glen, John. *Early Scottish Melodies.* Edinburgh: J. and R. Glen, 1900.

Gombrich, Ernst Hans. *Art and Illusion.* Princeton, NJ: Princeton University Press, [1960] 2000.

Goodwin, W. L. "The Nature of Music and the Music of Nature." *Queen's Quarterly* 11, no. 2 (Oct. 1903): 193–202.

Grainger, Percy. "The Impress of Personality in Unwritten Music." *Musical Quarterly* 1, no. 3 (July 1915): 416–35.

———. "Collecting with the Phonograph." *Journal of the Folk-Song Society* 3, no. 12 (1908): 147–62.

Grout, Donald Jay. *A History of Western Music.* New York: Norton, 1960.

———. *A Short History of Opera.* Vol. 1. New York: Columbia University Press, 1947.

Gurney, Edmund. *The Power of Sound.* London: Smith, Elder, 1880.

———. "On Some Disputed Points in Music." *Fortnightly Review* 20, no. 115 (July 1876): 106–30.

Guthrie, William Norman. "A New Star." *Sewannee Review* 12, no. 3 (July 1904): 354–60.

Hagar, George J., ed. *The New Supreme Webster Dictionary.* New York: Adair and Petty, 1919.

Haldane, J. B. S. "Differences." *Mind* 57, no. 227 (July 1948): 294–301.

Ham, Cody J. "How Does Your Voice Look Sent Over the Telephone?" *Telephony* 64 (Jan.–July, 1913): 92.

Harman, Alec. "Review: One-Volume History." *Musical Times* 109, no. 1438 (Dec. 1962): 845, 847.

Hawkins, Chauncey J. "Sexual Selection and Bird Song." *Auk* 35, no. 4 (Oct. 1918): 421–37.

Haydon, Glen. *Introduction to Musicology: A Survey of the Fields, Systematic and Historical, of Musical Knowledge and Research.* Westport, CT: Greenwood, [1941] 1978.

Helmholtz, Hermann L. F. *On the Sensations of Tone as a Physiological Basis for the Theory of Music.* Translated by Alexander J. Ellis. London: Longman, Green, [1863] 1875.

Hemingway, Ernest. *The Garden of Eden.* New York: Scribner, 1987.

Herzog, George. "Do Animals Have Music?" *Bulletin of the American Musicological Society* 5 (Aug. 1941): 3–4.

Hill, Peter. "From Réveil des oiseaux to Catalogue d'oiseaux: Messiaen's Cahiers de notations des chants d'oiseaux, 1952–59." In *Messiaen Perspectives I: Sources and Influences,* ed. Christopher Dingle and Robert Fallon, 143–74. Burlington, VT: Ashgate, 2013.

Hill, Peter, and Nigel Simeone. *Oiseaux Exotiques.* Burlington, VT: Ashgate, 2007.

———. *Messiaen.* New Haven: Yale University Press, 2005.

Hinde, R. A. "William Homan Thorpe." *Biographical Memoirs of Fellows of the Royal Society* 33 (Dec. 1987): 621–39.

Hipsher, Edward Ellsworth. "Do Animals Like Music?" *Musical Quarterly* 12, no. 2 (Apr. 1926): 166–74.

Hold, Trevor. "Messiaen's Birds." *Music and Letters* 52, no. 2 (Apr. 1971): 113–22.

———. "The Notation of Bird-Song: A Review and a Recommendation." *Ibis* 112, no. 2 (1970): 151–72.

Hornbostel, Erich M. von. *Tonart und Ethos: Aufsätze zur Musikethnologie und Musik-psychologie.* Edited by Christian Kaden and Erich Stockmann. Leipzig: Reclam, 1986.

———. "Formanalysen an siamesischen Orchesterstücken." *Archiv für Musikwissenschaft* 2, no. 2 (Apr. 1920): 306–33.

Hornbostel, Erich M. von, and Curt Sachs. "Classification of Musical Instruments." Translated by Anthony Baines and Klaus P. Wachsmann. *Galpin Society Journal* 14 (Mar. 1961): 3–29.

Hunt, Richard. "Evidence of Musical 'Taste' in the Brown Towhee." *Condor* 24, no. 6 (Nov.–Dec. 1922): 193–203.

Hunter, George William. *Elements of Biology.* New York: American Book, 1907.

Huplitz, Agnes. "Poliomyelitis and Cranberries." *American Journal of Nursing* 17, no. 6 (Mar. 1917): 502–4.

Huxley, Julian. *Evolution: The Modern Synthesis.* 2nd ed. London: Allen and Unwin, [1942] 1963.

Huxley, Thomas Henry. *Introduction to the Study of Zoology: The Crayfish*. 4th ed. London: Kegan Paul, Trench, 1880.

Ingraham, Sydney E. "Instinctive Music." *Auk* 55, no. 4 (Oct. 1938): 614–28.

Jackendoff, Ray, and Fred Lerdahl. *A Generative Theory of Tonal Music*. Cambridge, MA: MIT Press, 1981.

Jairazbhoy, Nazir A. "The 'Objective' and Subjective View in Music Transcription." *Ethnomusicology* 21, no. 2 (May 1977): 263–73.

Jameson, Fredric. *Postmodernism, or, The Cultural Logic of Late Capitalism*. Durham, NC: Duke University Press, 1991.

Jastrow, Joseph. "Experimental Psychology in Leipzig." *Science* 8, no. 198 (Nov. 1886): 459–62.

Kalikow, Theodora J., and John A. Mills. "Wallace Craig (1876–1954), Ethologist and Animal Psychologist." *Journal of Comparative Psychology* 103, no. 3 (1989): 281–88.

Kames, Henry Home, Lord. *Elements of Criticism*. Edited by Abraham Mills. New York: Huntington and Mason, 1855.

Kearney, Thomas H., and Robert H. Peebles. *Flowering Plants and Flowers of Arizona*. Washington, DC: United States Department of Agriculture, 1942.

Kircher, Athanasius. *Musurgia universalis sive ars magna consoni et dissoni, in X libros digesta*. Vol. 1. Rome: Francesco Corbelletti, 1650.

Klotz, Sebastian. "Hornbostel, Erich Moritz." In *Die Musik in Geschichte und Gegenwart: allgemeine Enzyklopädie der Musik*, ed. Ludwig Finscher, 356–64. Kassel: Bärenreiter, 2003.

Krause, Bernie. *Wild Soundscapes: Discovering the Voice of the Natural World*. Berkeley: Wilderness, 2002.

Krogman, Wilton Marion. "Physical Anthropology and Race Relations: A Biosocial Evaluation." *Scientific Monthly* 66, no. 4 (Apr. 1948): 317–21.

Krohn, William. "Facilities in Experimental Psychology at the Various German Universities." *American Journal of Psychology* 4, no. 4 (Aug. 1892): 585–94.

Kroodsma, Donald. *The Backyard Birdsong Guide: Eastern and Central North America*. San Francisco: Chronicle, 2008.

———. *The Backyard Birdsong Guide: Western North America*. San Francisco: Chronicle, 2008.

———. *The Singing Life of Birds*. New York: Houghton Mifflin, 2005.

Kroodsma, Donald, and Noah K. Strycker. "A *Birding* Interview with Donald Kroodsma." *Birding* 41, no. 3 (May 2009): 18–21.

Kunst, Jaap. "Zum Tode Erich von Hornbostel's." *Anthropos* 32, no. 1/2 (Jan.–Apr. 1937): 239–46.

———. "A Musicological Argument for Cultural Relationship between Indonesia: Probably the Isle of Java, and Central Africa." *Proceedings of the Musical Association* 62

(1936): 57–76.

Lach, Robert. *Studien zur Entwickelungsgeschichte der ornamentalen Melopöie*. Lepizig: Kahnt Nachfolger, 1913.

Lahy-Hollebeque, M., ed. *L'évolution humaine des origines à nos jours*. Vol. 3. Paris: Librairie Aristide Quillet, 1934.

Lang, Paul. "Music Scholarship at the Crossroads." *Musical Quarterly* 31, no. 3 (July 1945): 371–80.

Langendorff, Oskar. *Physiologische Graphik: Ein Leitfaden der in der Physiologie Gebräuchlichen Registrirmethoden*. Leipzig: Franz Deuticke, 1891.

Langfield, Herbert S. "Stumpf's 'Introduction to Psychology.'" *American Journal of Psychology* 50, no. 1/4 (Nov. 1937): 33–56.

LaRue, Jan. *Guidelines for Style Analysis*. New York: Norton, 1970.

Lawson, Robert W., and Richard K. Walton. *Birding by Ear: Eastern and Central North America*. New York: Houghton Mifflin, 2002. Compact disc set.

Lees, Frederic. "Photographing Speech." Photographs by Laurence, Paris. *Windsor Magazine* 27 (London: Ward, Lock, Jan. 1908): 257–64.

Lévi-Strauss, Claude. *The Raw and the Cooked: Mythologiques*. Vol. 1. Translated by John and Doreen Weightman. Chicago: University of Chicago Press, 1983.

———. *The Raw and the Cooked*. New York: Harper and Row, 1969.

Lewis, C. S. *The Magician's Nephew*. New York: Harper Collins, 1994.

*Library of Congress Strategic Plan 2011–2016*. Washington, DC: Library of Congress, Nov. 2010.

Lineff, Eugenie. *The Peasant Songs of Great Russia*. Second series. Moscow: Municipal Printing Office, 1912.

List, George. *Stability and Variation in Hopi Song*. Philadelphia: American Philosophical Society, 1993.

———. "The Reliability of Transcription." *Ethnomusicology* 18, no. 3 (Sept. 1974): 353–77.

———. "The Musical Significance of Transcription (Comments on Hood, 'Musical Significance')." *Ethnomusicology* 7, no. 3 (Sept. 1963): 194–95.

Lockwood, Samuel. "A Singing Hesperomys." *American Naturalist* 5, no. 12 (Dec. 1871): 761–70.

Lohmann, Michael, and Jean C. Roché. *Guide des Oiseaux Chanteurs*. Translated by Monique Schleiss. Aartselaar: Chantecler, 2005.

Lomax, Alan. "Folk Song Style: Notes on a Systematic Approach to the Study of Folk Song." *Journal of the International Folk Music Council* 8 (1956): 48–50.

Lomax, Alan, and Norman Berkowitz. "The Evolutionary Taxonomy of Culture." *Science* 177, no. 4045 (July 21, 1972): 228–39.

Luper, Albert T. "Review: A History of Western Music by Donald Jay Grout." *Notes* 18, no. 1 (Dec. 1960): 47–48.

Lyotard, Jean-François. *The Postmodern Condition: A Report on Knowledge*. Translated by Geoff Bennington and Brian Massumi. Minneapolis: University of Minnesota Press, 1984.

MacLeod, Julius. *The Quantitative Method in Biology*. New York: Longman, Green, 1919.

*Maine, A Guide "Down East."* State of Maine WPA, Federal Writers' Project, 1937.

Maine, Basil. "Shaw, Wells, Binyon—And Music." *Musical Quarterly* 18, no. 3 (July 1932): 375–82.

Marler, Peter. "Ethology and the Origins of Behavioral Endocrinology." *Hormones and Behavior* 47 (2005): 493–502.

Marler, Peter, and Hans Slabbekoorn. *Nature's Music: The Science of Birdsong*. New York: Elsevier, 2004.

Mason, Daniel Gregory. *The Art of Music*. Vol. 1. New York: National Society of Music, 1915.

Mathews, Ferdinand Schuyler. *Field Book of Wild Birds and Their Music*. New York: Putnam, [1904] 1921.

Mayr, Ernst. "Review: Origin of the Human Races." *Science*, new series 138, no. 3538 (Oct. 19, 1962): 422.

Mayr, Ernst, and William B. Provine, eds. *The Evolutionary Synthesis: Perspectives on the Unification of Biology*. Cambridge, MA: Harvard University Press, 1980.

McCormack, T. J. "Review: *An Introduction to Comparative Psychology*." *Monist* 5, no. 3 (Apr. 1895): 443–49.

Mead, Margaret. *Continuities in Cultural Evolution*. New Brunswick, NJ: Transaction Publishers, [1964] 1999.

Merriam, Alan P. *The Anthropology of Music*. Evanston, IL: Northwestern University Press, 1964.

Merry, Glen. "Voice Inflection in Speech." *Psychological Monographs of the American Psychological Association* 31 (Princeton, NJ: Psychological Review, 1922): 205–29.

Messiaen, Olivier. *Catalogue d'oiseaux*. Paris: Leduc, 1964.

Messiaen, Olivier, and Claude Samuel. *Entretiens avec Olivier Messiaen*. Paris: Belfond, 1967.

Metfessel, Milton. "Strobophotography in Bird Singing." *Science* 79, no. 2053 (May 4, 1934): 412–13.

———. *Phonophotography in Folk Music: Negro Songs in New Notation*. Chapel Hill: University of North Carolina Press, 1928.

———. "The Collecting of Folk Songs by Phonophotography." *Science* 67, no. 1724 (Jan. 13, 1928): 28.

Miller, D. C. *Science of Musical Sounds*. New York: Macmillan, 1916.

Miller, Ray E. "A Strobophotographic Analysis of a Tlingit Indian's Speech." *International Journal of American Linguistics* 6, no. 1 (Mar. 1930): 47–68.

Moore, Robert. "The Fox Sparrow as a Songster." *Auk* 30, no. 2 (Apr. 1913): 177–87.

Morgan, Conwy Lloyd. *Animal Behaviour*. London: Edward Arnold, 1900.

———. *An Introduction to Comparative Psychology*. New York: Scribner, [1894] 1896.

Morrow, Edward. "Poor Old Uncle Tom." *Prairie Schooner* 4, no. 3 (Summer 1930): 174–80.

Munro, Thomas. *Evolution in the Arts, and other Theories of Culture History*. New York: Abrams, 1963.

"Music for Animals." *Musical Times and Singing Class Circular* 38, no. 649 (Mar. 1, 1897): 162–63.

National Audubon Society. *Audubon Christmas Bird Count Compiler's Manual*. Willow Grove, PA: Audubon Science, 2014.

Oldys, Henry W. "A Remarkable Hermit Thrush Song," *Auk* 30, no. 4 (Oct. 1913): 538–41.

———. "Music of Man and Bird." *Harper's Monthly* 114 (Apr. 1907): 766–71.

———. "The Rhythmical Song of the Wood Pewee." *Auk* 21, no. 2 (Apr. 1904): 270–74.

Palmer, T. S. "Obituaries: Ferdinand Schuyler Mathews." *Auk* 57, no. 1 (Jan. 1940): 138.

Parry, Charles Hubert. *Style in Musical Art*. London: Macmillan, 1911.

———. *The Evolution of the Art of Music*. Edited by H. C. Colles. New York: Appleton, [1893] 1921.

Pavlov, Ivan. "New Researches on Conditioned Reflexes." *Science* 58, no. 1506 (Nov. 9, 1923): 359–61.

———. "The Scientific Investigation of the Psychical Faculties or Processes in the Higher Animals." *Science* 24, no. 620 (Nov. 16, 1906): 613–19.

Penot, Jacques. "Olivier Messiaen ornithologue." In *Portrait(s) d'Olivier Messiaen*, ed. Catherine Massip, 61–74. Paris: Bibliothèque Nationale de France, 1996.

Peterson, Roger Tory. *A Field Guide to the Birds of Eastern and Central North America*. Boston: Houghton Mifflin, 1980.

Potter, Ralph K. "Audivisual Music." *Hollywood Quarterly* 3, no. 1 (Autumn 1947): 66–78.

———. "Visible Patterns of Sound." *Science* 102, no. 2654 (Nov. 9 1945): 463–70.

*Prohibiting Vivisection of Dogs: Hearings Before the Subcommittee of the Committee on the Judiciary United States Senate*. Washington, DC: United States Government Printing Office, 1919.

"Review: An Introduction to Comparative Psychology." *British Medical Journal* 1, no. 1788 (Apr. 1895): 761–62.

"Review: The Power of Sound, by Edmund Gurney." *British Quarterly Review* 73, no. 145 (Jan. 1881): 209–10.

Riddle, Oscar. "Genetics and Human Behavior." *American Biology Teacher* 10, no. 3 (Mar. 1948): 69–74.

Robbins, C. A. "The Identity of Cladonia Beaumontii." *Rhodora* 29, no. 343 (July 1927): 133–38.

Roberts, Helen. "New Phases in the Study of Primitive Music." *American Anthropologist* 24, no. 2 (Apr.–June 1922): 144–60.

Roe, Edward T., ed. *Laird and Lee's Webster's New Standard American Dictionary*. Chicago: Laird and Lee, 1911.

Romanes, George J. "Singing Mice." *Nature* 17, no. 418 (Nov. 8, 1877): 29.

Rood, Louise. "An Invaluable Survey of Western Music." *Massachusetts Review* 2, no. 1 (Autumn 1960): 179–81.

Rosen, Charles. *The Classical Style: Haydn, Mozart, Beethoven*. New York: Norton, 1971.

Sachs, Curt. "La signification, la tâche et la technique muséographique des collections d'instruments de musique." Reprinted from *Mouseion* 27/28, in *Cahiers de musiques traditionelles* 16 (2003): 11–41.

——. *The Wellsprings of Music*. New York: Da Capo, 1962.

Sandeman, George. *Problems of Biology*. London: Swan Sonnenschein, 1896.

Sanford, Edmund C. "A Laboratory Course in Physiological Psychology. (Third Paper): V. Vision." *American Journal of Psychology* 4, no. 3 (Apr. 1892): 474–90.

Sanford, Samuel N. F. *New England Herbs, Their Preparation and Use*. Boston: New England Museum of Natural History, 1937.

Santayana, George. *The Life of Reason, or The Phases of Human Progress: Reason in Common Sense*. New York: Scribner, [1905] 1920.

Saunders, Aretas A. "The Song of the Song Sparrow." *Wilson Bulletin* 63, no. 2 (June 1951): 99–109.

——. "Review: The Song of the Wood Pewee." *Auk* 61, no. 4 (Oct. 1944): 658–60.

——. "The Song of the Field Sparrow." *Auk* 39, no. 3 (July 1922): 386–99.

——. "Evolution of Bird Song." *Auk* 36, no. 1 (Jan. 1919): 149–51.

——. "Some Suggestions for Better Methods of Recording and Studying Bird Songs." *Auk* 32, no. 2 (Apr. 1915): 173–83.

Scarborough, Deborah. *A Song Catcher in Southern Mountains: American Folk Songs of British Ancestry*. New York: Columbia University Press, 1937.

Schafer, R. Murray. *The Soundscape: Our Sonic Environment and the Tuning of the World*. Rochester, VT: Destiny, [1977] 1994.

Schoen, Max. "An Experimental Study of the Pitch Factor in Artistic Singing." *Psychological Monographs of the American Psychological Association* 31 (Princeton, NJ: Psychological Review, 1922): 230–59.

Schuchert, Charles, and S. S. Buckman. "The Nomenclature of Types in Natural History." *Science* 21, no. 545 (June 9, 1905): 899–901.

"The Scientific Basis for Human Unity." *Unesco Courier* 3, no. 6 (July 1950): 8–9.

Scott, Edouard-Léon. *The Phonographic Manuscripts of Edouard-Léon Scott de Martinville*. Edited and translated by Patrick Feaster. Bloomington, IN: First Sounds, 2009.

Seashore, Carl E. "Psychology in Music: The Rôle of Experimental Psychology in the Science and Art of Music." *Musical Quarterly* 16, no. 2 (Apr. 1930): 229–37.

——. "Phonophotography in the Measurement of the Expression of Emotion in Music and Speech." *Scientific Monthly* 24, no. 5 (May 1927): 463–71.

Seeger, Charles. "Prescriptive and Descriptive Music-Writing." *Musical Quarterly* 44, no. 2 (Apr. 1958): 184–95.

———. "Toward a Universal Music Sound-Writing for Musicology." *Journal of the International Folk Music Council* 9 (1957): 63–66.

———. "An Instantaneous Music Notator." *Journal of the International Folk Music Council* 3 (1951): 103–6.

Sharp, Cecil James. "Folk-Song Collecting." *Musical Times* 48, no. 767 (Jan. 1907): 16–17.

Sharp, Cecil James, and Charles L. Marson. *Folk Songs from Somerset.* London: Simpkin, 1905.

Shelemay, Kay Kaufman. *Soundscapes: Exploring Music in a Changing World.* New York: Norton, 2001.

Sherlaw-Johnson, Robert. "Birdsong." In *The Messiaen Companion*, ed. Peter Hill, 249–65. Portland, OR: Amadeus, 1995.

———. *Messiaen.* Berkeley: University of California Press, [1975] 1989.

Sidebotham, Joseph. "Singing Mice." *Nature* 17, no. 418 (Nov. 8, 1877): 29.

Slater, Peter. "Fifty Years of Bird Song Research: A Case Study in Animal Behavior." *Animal Behavior* 65, no. 4 (2003): 633–39.

Smith, Harlan I. "Museums of Sounds." *Science* 40, no. 1025 (Aug. 1914): 273–74.

Sonneck, Oscar G. "Music for Adults and Music for Children." *Music Supervisors' Journal* 15, no. 2 (Dec. 1928): 21, 23, 25, 27.

Spearman, Charles E. "The Confusion That Is Gestalt-Psychology." *American Journal of Psychology* 50, no. 1 (Nov. 1937): 369–83.

Spencer, Herbert. "The Origin and Function of Music." In *Essays: Scientific, Political, and Speculative*, vol. 2. New York: Appleton, 1907.

———. *The Study of Sociology.* New York: Appleton, [1878] 1901.

———. "The Origin of Music." *Mind* 15, no. 60 (Oct. 1890): 29, 425–26.

"Status Approuvés par Décision Minisérielle du 8 Mars 1866." In *Bulletin de la Société de Linguistique de Paris*, vol. 1. Paris: 1871.

Steward, Julian. *Theory of Culture Change: The Methodology of Multilinear Evolution.* Champaign: University of Illinois Press, [1955] 1972.

Stratton-Porter, Gene. *Moths of the Limberlost.* New York: Doubleday, 1916.

———. *Music of the Wild.* Cincinnati: Jennings and Graham, 1910.

Stumpf, Carl. *Die Anfange der Musik.* Lepizig: Barth, 1911.

———. "Das Psychologische Institut." In G. Lenz (Hrsg.), *Geschichte der königlichen Friedrich-Wilhelm Universität*, vol. 3, 202–7. Halle: Verlag der Buchhandlung des Waisenhauses, 1910.

Stumpf, Carl, and E. M. von Hornbostel. "Über die Bedeutung ethnologischer Untersuchungen für die Psychologie und Ästhetik der Tonkunst." *Beiträge zur Akustik und*

*Musikwissenschaft* 6 (1911): 102–15.

Sturvetant, A. H. "Social Implications of the Genetics of Man." *Science* 120, no. 3115 (Sept. 1954): 405–7.

Sully, James. "Animal Music." *Cornhill* 40 (Nov. 1879): 605–21.

Teale, Edwin. "Test-tube Magic Creates Amazing New Flowers." *Popular Science* (May 1940): 57–64.

Thimonnier, René. "Principes d'une Propagande musicale française en Allemagne occupé." *La Revue Musicale* 201 (Sept. 1946): 313–14.

Thomas, Roy. *Stan Lee's Amazing Marvel Universe*. New York: Sterling, 2006.

Thone, Frank. "Not 'Stuffed.'" *Science News-Letter* 28, no. 752 (Sept. 7, 1935): 154–56.

Thorpe, William Homan. "The Biological Significance of Duetting and Antiphonal Song." *Acta Neurobiologiae Experimentalis* 35, no. 5–6 (1975): 517–28.

———. "Is There a Comparative Psychology? The Relevance of Acquired Constraints in the Action Patterns and Perceptions of Animals." *Annals of the New York Academy of Sciences* 223 (Dec. 1973): 89–112.

———. *Duetting and Antiphonal Song in Birds: Its Extent and Significance*. Leiden: Brill, 1972.

———. "Comparative Psychology." *Annual Review of Psychology* 12 (1961): 27–50.

———. "The Learning of Song Patterns by Birds, with Especial Reference to the Song of the Chaffinch Fingilla Coelebs." *Ibis* 100, no. 4 (1958): 535–70.

———. "Some Implications of the Study of Animal Behavior." *Scientific Monthly* 84, no. 6 (June 1957): 309–20.

———. "The Process of Song-Learning in the Chaffinch as Studied by Means of the Sound Spectrograph." *Nature* 173, no. 4402 (Mar. 13, 1954): 465–69.

Thurston, Harry, ed. *New Websterian 1912 Dictionary*. New York: Syndicate, 1912.

Treutler, W. J. "Music in Relation to Man and Animals." *Proceedings of the Musical Association*, 25th Session (1899): 71–91.

Tuan, Yi-Fu. "Sign and Metaphor." *Annals of the Association of American Geographers* 68, no. 3 (Sept. 1978): 363–72.

*Unesco Courier*. Vols. 1–3. Paris: UNESCO, 1950.

Vernon, E. "The Apprehension and Cognition of Music." *Proceedings of the Musical Association*, 59th Sess. (1932–33): 61–64.

Virchow, Rudolf. "Über den Werth des pathologischen Experiments." *Transactions of the International Medical Congress*, 1 (London: Kolckmann, 1881): 22–37.

"The Vivisection Question in Germany." *North Carolina Medical Journal* 4, no. 1 (July 1879): 36–40.

Voegelin, Eric. "The Growth of the Race Idea." *Review of Politics* 2, no. 3 (July 1940): 283–317.

Voigt, Alwin. *Exkursionsbuch zum Studium der Vogelstimmen: Praktisch Anleitung zum*

*Bestimmen der Vögel nach irhem Gesange.* Leipzig: Quelle and Meyer, 1909.

W., J. A. "Review: *A History of Western Music* by Donald Jay Grout." *Music and Letters* 43, no. 4 (Oct. 1962): 363–65.

Wallace, Alfred Russel. *Darwinism: An Exposition of the Theory of Natural Selection.* New York: Macmillan, 1889.

Wallace, Alfred Russel, and James Marchant. *Letters and Reminiscences.* New York: Harper, 1916.

Wallace, William. *The Threshold of Music.* New York: Macmillan, 1908.

Wallaschek, Richard. *Primitive Music.* New York: Longman, Green, 1893.

Wallaschek, Richard, and James McKeen Cattell. "On the Origin of Music." *Mind* 16, no. 63 (July 1891): 375–88.

Warren, Katherine Brehme, ed. *Origin and Evolution of Man.* Cold Spring Harbor Symposia on Quantitative Biology XV. Cold Spring Harbor, NY: Biological Laboratory at Cold Spring Harbor, 1951.

Wead, Charles K. "The Study of Primitive Music." *American Anthropologist* 2, no. 1 (Jan. 1900): 75–79.

*Webster's Collegiate Dictionary.* 3rd ed. Springfield, MA: Merriam, 1916.

Weismann, August. "The Musical Sense in Animals and Men." *Popular Science Monthly* 37 (July 1890): 352–58.

White, Leslie. *The Evolution of Culture: The Development of Civilization to the Fall of Rome.* Walnut Creek, CA: Left Coast Press, [1959] 2007.

Whitman, Charles Otis. *Inheritance, Fertility, and the Dominance of Sex and Color in Hybrids of Wild Species of Pigeons: Posthumous Works of Charles Otis Whitman.* Edited by Oscar Riddle. Washington, DC: Carnegie Institute, 1919.

Wilson, Guy West. "The Identity of Mucor Mucedo." *Bulletin of the Torrey Botanical Club* 33, no. 11 (Nov. 1906): 557–60.

Witchell, Charles A. *The Evolution of Bird-Song, with Observations on the Influence of Heredity and Imitation.* London: Adam and Charles Black, 1896.

Wolters, A. W. "Review: Principles of Gestalt Psychology by. K. Koffka." *Philosophy* 11, no. 44 (Oct. 1936): 502–4.

## Secondary Sources

Ames, Eric. "The Sound of Evolution." *Modernism/modernity* 10, no. 2 (Apr. 2003): 297–325.

Amundson, Ronald. *The Changing Role of the Embryo in Evolutionary Thought.* New York: Cambridge University Press, 2005.

Ash, Mitchell G. *Gestalt Psychology in German Culture 1890–1967: Holism and the Quest for Objectivity.* New York: Cambridge University Press, 1998.

———. "Academic Politics in the History of Science: Experimental Psychology in Germany, 1879–1941." *Central European History* 13, no. 3 (Sept. 1980): 272–73.

Auslander, Philip. *Liveness: Performance in a Mediatized Culture*. New York: Routledge, 1999.

Bal, Mieke. "Telling, Showing, Showing Off." *Critical Inquiry* 18, no. 3 (Spring 1992): 556–94.

Balibar, Etienne. *Identity and Difference: John Locke and the Invention of Consciousness*. Translated by Warren Montag. New York: Verso, 2013.

———. "Citizen Subject." In *Who Comes After the Subject?*, ed. Eduardo Cadava, Peter Connor, and Jean-Luc Nancy, 33–57. New York: Routledge, 1991.

Barrow, Mark V. *Nature's Ghosts: Confronting Extinction from the Age of Jefferson to the Age of Ecology*. Chicago: University of Chicago Press, 2009.

Bartók, Peter. *My Father*. Tampa, FL: Bartók Records, 2002.

Barz, Gregory F., and Timothy J. Cooley, eds. *Shadows in the Field: New Perspectives for Fieldwork in Ethnomusicology*. 2nd ed. New York: Oxford University Press, 2008.

Behar, Ruth, and Deborah A. Gordon, eds. *Women Writing Culture*. Berkeley: University of California Press, 1995.

Bennett, Jane. *Vibrant Matter: A Political Ecology of Things*. Durham, NC: Duke University Press, 2010.

Bowler, Peter. *Evolution: The History of an Idea*. Berkeley: University of California Press, 1989.

Brady, Erika. *A Spiral Way: How the Phonograph Changed Ethnography*. Jackson: University of Mississippi Press, 1999.

Braidotti, Rosi. *The Posthuman*. Malden, MA: Polity, 2013.

Brown, Gwynne Kuhner. "Problems of Race and Genre in the Critical Reception of Porgy and Bess." PhD diss. University of Washington, 2006.

Brown, Julie. *Bartók and the Grotesque: Studies in Modernity, the Body, and Contradiction in Music*. Burlington, VT: Ashgate, 2007.

Bruyninckx, Joeri. "Sound Science: Recording and Listening in the Biology of Bird Song, 1880–1980." PhD diss. University of Maastricht, 2012.

———. "Sound Sterile: Making Scientific Field Recordings in Ornithology." In *The Oxford Handbook of Sound Studies*, ed. Karin Bijsterveld and Trevor Pinch, 127–50. New York: Oxford University Press, 2011.

Burch, Susan. *Signs of Resistance: American Deaf Cultural History, 1900 to World War II*. New York: New York University Press, 2002.

Burkhardt, Richard. *Patterns of Behavior: Konrad Lorenz, Niko Tinbergen, and the Founding of Ethology*. Chicago: University of Chicago Press, 2005.

———. "Constructing the Zoo: Science, Society, and Animal Nature at the Paris Menagerie 1794–1883." In *Animals in Human Histories: The Mirror of Nature and Culture*, ed. Mary Henniger-Voss, 231–57. Rochester, NY: University of Rochester Press, 2002.

Burnett, D. Graham. *The Sounding of the Whale: Science and Cetaceans in the Twentieth Century*. Chicago: University of Chicago Press, 2012.

Butler, Judith. *Senses of the Subject*. New York: Fordham University Press, 2015.

Carroll, Mark. *Music and Ideology in Cold War Europe*. New York: Cambridge University Press, 2003.

Chansigaud, Valérie. *Histoire de l'ornithologie*. Paris: Delachaux et Niestlé, 2007.

Chaudhuri, Una. "(De)Facing the Animals: Zooësis and Performance." *Drama Review* 51, no. 1 (Spring 2007): 8–20.

Chaudhuri, Una, and Holly Hughes, eds. *Animal Acts: Performing Species Today*. Ann Arbor: University of Michigan Press, 2014.

Chen, Mel Y. *Animacies: Biopolitics, Racial Mattering, and Queer Affect*. Durham, NC: Duke University Press, 2012.

Cheong, Wai-Ling. "Messiaen's Triadic Coloration: Modes as Intervention." *Music Analysis* 21 (2002): 53–84.

Cixous, Hélène. "The Laugh of the Medusa." Translated by Keith Cohen and Paula Cohen. *Signs* 1, no. 4 (Summer 1976): 875–93.

Clark, Suzannah, and Alexander Rehding, eds. *Music Theory and Natural Order from the Renaissance to the Early Twentieth Century*. New York: Cambridge University Press, 2001.

Comfort, Nathaniel. *The Science of Human Perfection: How Genes Became the Heart of American Medicine*. New Haven: Yale University Press, 2012.

———. *The Tangled Field: Barbara McClintock's Search for the Patterns of Genetic Control*. Cambridge, MA: Harvard University Press, 2003.

Crist, Elizabeth B. "Aaron Copland and the Popular Front." *Journal of the American Musicological Society* 56, no. 2 (Summer 2002): 409–65.

Crutzen, Paul J., and Eugene F. Stoermer. "The Anthropocene." *Global Change Newsletter* 41 (2000): 17–18.

Cusick, Suzanne. *Francesca Caccini at the Medici Court: Music and the Circulation of Power*. Chicago: University of Chicago Press, 2009.

Daston, Lorraine. "Type Specimens and Scientific Memory." *Critical Inquiry* 31, no. 1 (Autumn 2004): 153–82.

Daston, Lorraine, and Peter Galison. *Objectivity*. New York: Zone, 2010.

Davis, Lennard J., ed. *The Disability Studies Reader*. 4th ed. New York: Routledge, 2013.

Dayan, Colin. *With Dogs at the Edge of Life*. New York: Columbia University Press, 2016.

Delumeau, Jean. *History of Paradise: The Garden of Eden in Myth and Tradition*. Translated by Matthew O'Connell. New York: Continuum, 2000.

Denning, Michael. *Noise Uprising: The Audiopolitics of a World Musical Revolution*. New York: Verso, 2015.

Derrida, Jacques. "Différance." Translated by Alan Bass. In *Margins of Philosophy*, 3–27.

Chicago: University of Chicago Press, 1982.

Dunlap, Thomas R. *In the Field, Among the Feathered: A History of Birders and Their Guides*. Oxford: Oxford University Press, 2011.

———. "Tom Dunlap on Early Bird Guides." *Environmental History* 10, no. 1 (2005): 110–18.

Eigen, Edward. "Overcoming First Impressions: Georges Cuvier's Types." *Journal of the History of Biology* 30, no. 2 (Summer 1997): 179–209.

Eley, Craig. "A Birdlike Act: Sound Recording, Nature Imitation, and Performance Whistling." *Velvet Light Trap* 74 (Fall 2014): 1–15.

Elliott, Clark A., and Margaret W. Rossiter. *Science at Harvard University: Historical Perspectives*. Cranbury, NJ: Associated University Presses, 1992.

Elston, Mary Ann. "Women and Anti-Vivisection in Victorian England, 1870–1900." In *Vivisection in Historical Perspective*, ed. Nicolaas A. Rupke, 259–95. New York: Croom Helm, 1987.

Erskine, Fiona. "The *Origin of Species* and the Science of Female Inferiority." In *Charles Darwin's The Origin of Species: New Interdisciplinary Essays*, ed. David Amigoni and Jeff Wallace, 95–121. New York: Manchester University Press, 1995.

Fallon, Robert. "The Record of Realism in Messiaen's Bird Style." In *Olivier Messiaen: Music, Art, and Literature*, ed. Christopher Dingle and Nigel Simeone, 115–36. Burlington, VT: Ashgate, 2007.

———. "Messiaen's Mimesis: The Language and Culture of the Bird Styles." PhD diss. University of California–Berkeley, 2005.

Farnum, Richard. "Patterns of Upper-Class Education in Four American Cities: 1875–1975." In *The High Status Track: Studies of Elite Schools and Stratification*, ed. Paul William Kingston and Lionel S. Lewis, 53–74. Albany: State University of New York, 1990.

Finney, John. *Music Education in England, 1950–2010: The Child-Centered Progressive Tradition*. Burlington, VT: Ashgate, 2011.

Foucault, Michel. *The Order of Things: An Archaeology of the Human Sciences*. Translated by Alan Sheridan. New York: Vintage, 1994.

Francis, Mark. *Herbert Spencer and the Invention of Modern Life*. New York: Routledge, 2007.

Fulcher, Jane. "Vincent d'Indy's 'Frame Anti-Juif' and Its Meaning in Paris, 1920." *Cambridge Opera Journal* 2, no. 3 (Nov. 1990): 295–319.

Garson, Marjorie. *Moral Taste: Aesthetics, Subjectivity and Social Power*. Buffalo, NY: University of Toronto Press, 2007.

Gertner, Jon. *The Idea Factory: Bell Labs and the Great Age of American Innovation*. New York: Penguin, 2012.

Glass, Aaron. "Frozen Poses: Hamat'sa Dioramas, Recursive Representation, and the Making of a Kwakwaka'wakw Icon." In *Photograph, Anthropology, and History: Expanding the Frame*, ed. Elizabeth Edwards and Christopher Morton, 89–116. Burlington, VT: Ashgate, 2009.

Glass, Bentley. "Milislav Demerec 1895–1966." Washington, DC: National Academy of Sciences, 1971.

Grosz, Elizabeth. *The Nick of Time: Politics, Evolution, and the Untimely*. Durham, NC: Duke University Press, 2004.

Gruen, Lori, ed. *The Ethics of Captivity*. New York: Oxford University Press, 2014.

Grusin, Richard, ed. *Anthropocene Feminism*. Minneapolis: University of Minnesota Press, 2017.

Hamlin, Kimberly. *From Eve to Evolution: Darwin, Science, and Women's Rights in Gilded Age America*. Chicago: University of Chicago Press, 2014.

Haraway, Donna. *Primate Visions: Gender, Race, and Nature in the World of Modern Science*. New York: Routledge, 1989.

———. "Teddy Bear Patriarchy: Taxidermy in the Garden of Eden, New York City." *Social Text* 11 (Winter 1984–85): 20–64.

Harrington, Anne. *Reenchanted Science: Holism in German Culture from Wilhelm II to Hitler*. Princeton, NJ: Princeton University Press, 1996.

Harrison, Charles, Paul Wood, and Jason Gaiger, eds. *Art in Theory, 1815–1900: An Anthology of Changing Ideas*. Malden, MA: Blackwell, 1998.

Harrison, Peter. *The Bible, Protestantism, and the Rise of Natural Science*. New York: Cambridge University Press, 1998.

Hayles, Katherine. *How We Became Posthuman: Virtual Bodies in Cybernetics, Literature, and Informatics*. Chicago: University of Chicago Press, 1991.

Heywood, Linda Marinda. *Contested Power in Angola, 1840s to the Present*. Rochester, NY: University of Rochester Press, 2000.

Hicks, Michael. *Henry Cowell, Bohemian*. Champaign: University of Illinois Press, 2002.

Hui, Alexandra. *The Psychophysical Ear: Musical Experiments, Experimental Sounds, 1840–1910*. Cambridge, MA: MIT Press, 2013.

Ingold, Timothy. *Evolution and Social Life*. New York: Cambridge University Press, 1986.

Jackson, Zakiyyah Iman. "Outer Worlds: The Persistence of Race in Movement 'Beyond the Human.'" *GLQ* 21, 2–3 (June 2015): 215–18.

Kerman, Joseph, and Gary Tomlinson. *Listen*. 7th ed. Boston: Bedford/St. Martin's, 2012.

Kete, Kathleen. "Animals and Ideology." In *Representing Animals*, ed. Nigel Rothfels, 29–30. Bloomington: Indiana University Press, 2002.

Kevles, Daniel J. *Eugenics and the Uses of Human Heredity*. Berkeley: University of California Press, 1985.

Kim, Clare Jean. *Dangerous Crossings: Race, Species, and Nature in a Multicultural Age*. New York: Cambridge University Press, 2015.

Kistner, Alzada Carlisle. *An Affair with Africa: Expeditions and Adventures across a Continent.* Washington, DC: Island, 1998.

Ko, Aph, and Syl Ko. *Aphro-isms: Essays on Pop Culture, Feminism, and Black Veganism from Two Sisters.* New York: Lantern, 2017.

Köhler, Wolfgang. "Gespräche in Deutschland." Translated by Ann M. Hentschel. In *Physics and National Socialism: An Anthology of Primary Sources*, ed. Klaus Hentschel, 36–40. Germany: Birkhäuser, 1996.

Korshin, Paul J. *Typologies in England, 1650–1820.* Princeton, NJ: Princeton University Press, 1982.

Kursell, Julia. "A Gray Box: The Phonograph in Laboratory Experiments and Fieldwork, 1900–1920." In *The Oxford Handbook of Sound Studies*, ed. Karin Bijsterveld and Trevor Pinch, 176–97. New York: Oxford University Press, 2012.

Larson, Edward John. *Evolution: The Remarkable History of a Scientific Theory.* New York: Modern Library, 2004.

Laruelle, François. *General Theory of Victims.* Translated by Jessie Hock and Alex Dubilet. Malden, MA: Polity, 2015.

Latour, Bruno. *We Have Never Been Modern.* Translated by Catherine Porter. Cambridge, MA: Harvard University Press, 1993.

Latour, Bruno, and Steve Woolgar. *Laboratory Life: The Construction of Scientific Facts.* Los Angeles: Sage, 1979.

Law, John, and Michael Lynch. "Field Guides and the Descriptive Organization of Seeing: Birdwatching as an Exemplary Observational Activity." *Human Studies* 11, no. 2 (1988): 271–303.

Liang, Puping, Yanwen Xi, Xiya Zhang, Chenhui Ding, Rui Huang, Zhen Zhang, Jie Lv, Xiawei Xie, Yuxi Chen, et al. "CRISPR/Cas9-mediated gene editing in human tripronuclear zygotes." *Protein and Cell* 6, no. 5 (May 2015): 363–72.

Lippman, Edward. *A History of Western Musical Aesthetics.* Lincoln: University of Nebraska Press, 1992.

Macdonald, Helen. "'What Makes You a Scientist Is the Way You Look at Things': Ornithology and the Observer 1930–1955." *Studies in History and Philosophy of Biological and Biomedical Sciences* 33 (2002): 55–77.

Mills, Mara. "Deaf Jam: From Inscription to Reproduction to Information." *Social Text* 28, no. 1 (Spring 2010): 35–58.

Morin, Karen M. "Carceral Space: Prisoners and Animals." *Antipode* 48, no. 5 (Nov. 2016): 1317–36.

Morrow, Mary Sue. *German Music Criticism in the Late Eighteenth Century: Aesthetic Issues and Instrumental Music.* New York: Cambridge University Press, 1997.

Morton, Timothy. *Hyperobjects: Philosophy and Ecology After the End of the World.* Minneapolis: University of Minnesota Press, 2013.

Mundy, Rachel. "Museums of Sound: Audio Bird Guides and the Pleasures of Knowledge." *Sound Studies* 2 (2016). http://www.tandfonline.com/doi/abs/10.1080/20551940 .2016.1228842.

——. "Evolutionary Categories and Musical Style from Adler to America." *Journal of the American Musicological Society* 67, no. 3 (Fall 2014): 735–68.

——. "Nature's Music: Birds, Beasts, and Evolutionary Listening." PhD diss. New York University, 2010.

——. "Birdsong and the Image of Evolution." *Society and Animals* 17, no. 3 (2009): 206–23.

Muñoz, José Esteban. "Theorizing Queer Inhumanisms: The Sense of Brownness." *GLQ* 21, 2–3 (June 2015): 209–10.

——. *Cruising Utopia: The Then and There of Queer Futurity*. New York: New York University Press, 2009.

Myers, Helen, ed. *Ethnomusicology: Historical and Regional Studies*. New York: Norton, 1993.

Nettl, Bruno. *The Study of Ethnomusicology: Thirty-One Issues and Concepts*. Champaign: University of Illinois Press, 2005.

——. "Comparison and Comparative Method in Ethnomusicology." *Anuario Interamericano de Inverstigacion Musical* 9 (1973): 148–61.

Orwell, George. *Nineteen Eighty-Four*. New York: Signet Classics, 1977.

Pantalony, David. *Altered Sensations: Rudolph Koenig's Acoustical Workshop in Nineteenth-Century Paris*. New York: Springer, 2009.

Partridge, Linda Dugan. "By the Book: Audubon and the Tradition of Ornithological Illustration." *Huntington Library Quarterly* 59, no. 2/3 (1996): 269–301.

Pasler, Jann. "The Utility of Musical Instruments in the Racial and Colonial Agendas of Late Nineteenth-Century France." *Journal of the Royal Musical Association* 129, no. 1 (2004): 24–76.

Pescatello, Ann M. *Charles Seeger: A Life in American Music*. Pittsburgh: University of Pittsburgh Press, 1992.

Petit, Annie. "L'esprit de la science anglaise et les Français au XIXème siècle." *British Journal for the History of Science* 17, no. 3 (Nov. 1984): 273–93.

Pisani, Michael V. "'I'm an Indian Too': Creating Native American Identities in Nineteenth- and Early Twentieth-Century Music." In *The Exotic in Western Music*, ed. Johnathan Bellman, 218–57. Dexter, MI: Northwestern University Press, 1998.

Pluciennik, Mark. *Social Evolution*. London: Duckworth, 2005.

Povinelli, Elisabeth A. *Geontologies: A Requiem to Late Liberalism*. Durham, NC: Duke University Press, 2016.

Radick, Gregory. *The Simian Tongue: The Long Debate About Animal Language*. Chicago: University of Chicago Press, 2007.

Reardon, Jenny. *Race to the Finish: Identity and Governance in an Age of Genomics*. Princeton, NJ: Princeton University Press, 2005.

Rehding, Alexander. "The Quest for the Origins of Music in Germany Circa 1900." *Journal of the American Musicological Society* 53, no. 2 (Summer 2000): 345–85.

Rice, Timothy. *Ethnomusicology: A Very Short Introduction*. New York: Oxford University Press, 2014.

Riley, Glenda. *Women and Nature: Saving the "Wild" West*. Lincoln: University of Nebraska Press, 1999.

Roland-Manuel, Alexis. *Histoire de la Musique: des origines à Jean-Sébastien Bach*. Vol. 1. Paris: Éditions Gallimard, 1986.

Rothenberg, David. *Why Birds Sing: A Journey into the Mystery of Bird Song*. New York: Basic Books, 2006.

Sarosi, Bálint. "Hungary and Romania." In *Ethnomusicology: Historical and Regional Studies*, ed. Helen Myers, 187–96. New York: Norton, 1993.

Scafi, Alessandro. *Maps of Paradise*. Chicago: University of Chicago Press, 2014.

——— . *Mapping Paradise: A History of Heaven on Earth*. Chicago: University of Chicago Press, 2006.

Schuchat, Wilfred. *The Garden of Eden and the Struggle to Be Human: According to the Midrash Rabbah*. New York: Devora, 2006.

Scott, Joan Wallach. *Gender and the Politics of History*. New York: Columbia University Press, 1988.

Seshadri, Kalpana Rahita. *HumAnimal: Race, Law, Language*. Minneapolis: University of Minnesota Press, 2012.

Sinnott, Edmund W. *Albert Francis Blakeslee*. Washington, DC: National Academy of Sciences, 1959.

Sinnott, Edmund W., Amos G. Avery, Sophie Satina, and Jacob Rietsema. *Blakeslee: The Genus Datura*. New York: Ronald Press, 1959.

Smith, Woodruff. *Politics and the Sciences of Culture in Germany, 1840–1920*. New York: Oxford University Press, 1991.

Smocovitis, Vassilliki Betty. "Humanizing Evolution: Anthropology, the Evolutionary Synthesis, and the Prehistory of Biological Anthropology, 1927–1962." *Current Anthropology* 53, no. S5 (Apr. 2012): S108-25.

——— . "'It Ain't Over 'Til It's Over': Rethinking the Darwinian Revolution." *Journal of the History of Biology* 38, no. 1 (Spring 2005): 33–49.

——— . *The Evolutionary Synthesis and Evolutionary Biology*. Princeton, NJ: Princeton University Press, 1996.

——— . "The Evolutionary Synthesis and Evolutionary Biology." *Journal of the History of Biology* 25, no. 1 (Spring 1992): 1–65.

Somfai, Laszlo. "Analytical Notes on Bartók's Piano Year of 1926." *Studia Musicologica*

*Academiae Scientiarum Hungaricae* 26, Fasc. 1/4 (1984): 5–58.

Sprout, Leslie A. "The 1945 Stravinsky Debates: Nigg, Messiaen, and the Early Cold War in France." *Journal of Musicology* 26, no. 1 (Winter 2009): 85–131.

Stead, Mike, Sean Rorison, and Oscar Scafidi. *Angola*. Guildford, CT: Bradt Travel Guides, 2013.

Steege, Benjamin. *Helmholtz and the Modern Listener*. New York: Cambridge University Press, 2012.

Sterne, Jonathan. *The Audible Past: Cultural Origins of Sound Reproduction*. Durham, NC: Duke University Press, 2003.

Strasser, Bruno J. "Collecting Nature: Practices, Styles, and Narratives." *Osiris* 27, no. 1 (2012): 303–40.

Taylor, Hollis. *Is Birdsong Music? Outback Encounters with an Australian Butcherbird*. Bloomington: Indiana University Press, 2017.

Taylor, Timothy D. "Radio." In *Music, Sound, and Technology in America: A Documentary History of the Early Phonograph, Cinema, and Radio*, ed. Timothy D. Taylor, Mark Katz, and Tony Grajeda, 329–54. Durham, NC: Duke University Press, 2012.

Thacker, Toby. *Music After Hitler, 1945–1955*. Burlington, VT: Ashgate, 2007.

Tick, Judith. *Ruth Crawford Seeger: A Composer's Search for American Music*. New York: Oxford University Press, 1997.

Weheliye, Alexander. *Habeus Viscus: Racializing Assemblages, Biopolitics, and Black Feminist Theories of the Human*. Durham, NC: Duke University Press, 2014.

Wolfe, Cary. *What Is Posthumanism?* Minneapolis: University of Minnesota Press, 2010.

# INDEX

*Note*: Page numbers in *italics* refer to illustrations; page numbers with "n" or "nn" indicate endnotes.

Cold Spring Harbor, 66, 69, 74–75, 113, 114–21, 123–24

Comfort, Nathaniel, 115, 213n83

comparative psychology, 31–32, 71, 123–24, 131–35, 142

*The Condor* (Hunt), 37

Cornell University, 48, 54, 150–51, 163, 193n20

Cornish, Charles, 20–21, 28, 185n14

Craig, Wallace, 60–73, 80–83, 100, 140, 144–45

Crosby-Brown collection, 49, 191n3

Cullen, Thomas S., 86–87

cultural relativism, 110–11, 207n7

culture: animal, 67; and biology, 18, 22–24, 119, 124; and birdsong, 155–56; and empirical science, 89–91; and evolution, 13; and genetics, 113–14, 120, 131; human, 22, 32, 67, 147, 161–62, 168, 171; and musical vivisection, 98–99; and natural history, 42; and postmodern humanity, 109–11, 123–24, 152; and race, 120; and soundscapes, 166–67; and species, 177–78

Cusick, Suzanne, 61–63

Cuvier, Georges, 52–53

cylinders, 46–49, 92–95, 96, 113, 138, 140

Darwin, Charles, 17–19, 20, 22–32, 35–38, 67, 118

Darwin-Spencer debates, 17–19, 22–27, 185n5

Daston, Lorraine, 45, 127, 135–36

data, 92–97, 103

*Datura stramonium,* 114–18

Davenport, Charles, 4, 65–66, 69, 74, 113, 115–16, 119, 124, 137

deafness, 68–70

Delamain, Jacques, 158, 161

Delumeau, Jean, 153

Demerec, Milislav, 115–18, 122

Densmore, Frances, 50, 54, 100

*The Descent of Man* (Darwin), 20, 24–26

descriptive notation, 135, 215n35

difference: and animanities, 169–81; and aural identification, 166–67; biological, 67, 109; bodily, 61, 64–65, 73, 80–83; and

classification, 122, 175, 177; ethics of, 85, 100, 103–5, 143; and evolution, 27–28; and graphic representation, 100–101, 136; hearing, 146; and identification, 62–63, 175; and identity, 44–46, 174–75, 226n15; musical, 1–3, 51; overview, 7–10; and race, 61–63, 110–11, 115, 121–22; and relationship, 80–83; and song collection, 61–63, 80–83, 166–67; sonic, 61, 80; taxonomies of, 73, 174–75, 179; and typology, 56

digital audio guides. *See* audio guides

dioramas, 48, 52–53

disability, 61, 64–65, 68–73, 80–83, 198n12

Dobzhanzsky, Theodosius, 118–20

doublethink, 120–23, 211–12n72

doves, 67–68, 198–99n18

duetting, 217n68

Ebbinghaus, Hermann, 88

Edenic soundscapes, 152–54, 160, 165, 166–67, 179–80, 221n29

Edison, Thomas, 5–6, 93–95

Edison wax cylinders, 5–6, 47, 93–95

Eigen, Edward, 53

Ellis, Alexander, 98

empiricism, 89, 91, 113, 116, 144. *See also* knowledge

ergograph, 93

Etchécopar, Robert-Daniel, 158

ethnography, 47–50, 54, 74–75, 91

ethnology, 47

ethnomusicology, 113, 134, 142

ethology, 130–31

eugenics, 4, 66, 69, 74, 81, 115–17, 177–78, 181, 227n23

Eugenics Records Office, 66, 69, 74, 115–17, 122–24

evolution: biological, 22, 27, 211n67; of birdsong, 17–27; cultural, 13; and Darwin-Spencer debates, 17–19, 23–27; and difference, 27–28; and modern genetics, 115–18, 177; musical, 18–27, 38, 113, 147; and music's scientific history, 111–14; and race, 119–23, 165, 177; social, 27–30, 35–38, 188n54; and specimen collections, 52; and the synthesis movement, 110, 118–22, 207n8

*The Evolution of Bird-Song* (Witchell), 25

expeditions, 73–80

experimental psychology, 88–93, 98, 136–38, 142

experimental research, 74–75, 85–87

*The Expression of the Emotions in Man and Animals* (Darwin), 26, 67

Fallon, Robert, 158–59

Feld, Steven, 146–47, 154, 159–62, 166–67

Fewkes, Jesse Walter, 4–5, 47–50, 193n19

*Field Book of Wild Birds and Their Music* (Mathews), 2, 33, 34

field collection(s), 56–58

field guides, 146–53, 156, 161–64, 166

field recordings. *See* soundscapes

fieldwork, 73, 74, 142, 158, 194n38

Finney, John, 31

Fisher, Ronald, 116

folk songs, 30, 48, 55, 62, 77, 100, 197nn5–6

Foster, Stephen, 20, 33–35

Fuertes, Louis Agassiz, 149–51

Galison, Peter, 127, 135

garden. *See* Edenic soundscapes

Garner, Richard, 23

Garson, Marjorie, 31

Garstang, Walter, 128–29

Gestalt psychology, 130–32, 139, 144, 216nn52–53, 217n59

Gilman, Benjamin Ives, 50, 100

Ginsberg, Morris, 119

Glen, John, 185n15

Grainger, Percy, 48

graphic inscriptions, 5, 11, 46, 95, 100, 136

graphic notation, 50, 98–101, 102, 129, 132–33, 136, 216n48. *See also* transcription

"great man" theory, 28, 188n59

Grout, Donald, 111–12

Gurney, Edmund, 26

Haraway, Donna, 46, 53, 192n14

Hargreaves, Julia, 163

Harvard Museum of Comparative Zoology, 4–5, 65

Harvard University, 65, 68–69

Hawkins, Chauncey, 24, 37–38

hearing, 36–38, 68–70, 125–26, 128, 137–38

Helmholtz, Hermann von, 93

Hemenway Southwestern Archaeological Expedition, 47. *See also* cylinders

hermit thrush, 33–34, 34

Herzog, George, 19–20, 80, 141

Hespie the mouse, 20–21, 24

Hill, Peter, 157

Hinde, Robert, 132

*A History of Western Music* (Grout), 112

Hold, Trevor, 157

holotype. *See* type specimens

Hood, Mantle, 134–35

Hopi songs, 47, 49–50, 51, 135

Hornbostel, Erich Moritz von, 49–50, 84–85, 91–92, 95–105, 130–31, 133–38

Hui, Alexandra, 91, 92, 204n25

human bodies, 13, 58, 85, 144, 156, 176

humanism, 7, 13, 169, 171, 177–78, 180

humanity, 7–10, 23–24, 110, 118, 165

human listeners, 125–30, 138, 141

human song, 1–3, 17–19, 42–43, 142–43, 184n1

Hunt, Richard, 37

Huxley, Thomas, 44

*Ibis* (journal), 140

identification, 41–42, 45, 62–63, 68, 147, 149–51, 164, 166–67, 175

identity: and the animanities, 168–69, 173–74; biological, 44–46, 63, 67–68; categorical, 173–74; and classification, 45–46, 50–51; defined, 44–45; and difference, 44–46, 174–75, 226n15; in the field, 56; and language, 22–23; musical, 61–63, 81; natural, 146, 149, 166–67; overview, 42–44; and postmodern humanity, 109–11, 147–48, 152, 166–67; postwar, 123–24, 143, 178, 180; racial,

120–22; sonic, 152, 158–65; of species, 45, 68, 118; taxonomies of, 159, 173–75

independent research laboratories, 65–73

Ingold, Timothy, 27

Ingraham, Sydney, 100

institutional collections, 41–43, 46–50, 55, 56–58

institutional research support, 71, 73, 134

intelligent reflection, 85–88, 98–99, 102–5, 144, 176, 179

*An Introduction to Comparative Psychology* (Morgan), 31

Jairazbhoy, Nazir, 141

"The Keel Row," 20–21

"Kham hom," 84, 92, 97–98, 104

Kircher, Athanasius, 22

knowledge: of animals, 3–7; and the animanities, 175–76; ethics of, 85, 95, 126–27, 138, 145; musical, 58, 61–63, 92; natural, 41–43, 57, 166; objective and subjective, 135, 142, 144, 156, 171, 176, 178–79; quantitative, 104–5; value of, 103–4, 170, 176; and vivisection, 87, 95, 175–76

Kodály, Zoltan, 62

Koenig, Rudolph, 95, 136

Koffka, Kurt, 90, 131

Köhler, Wolfgang, 90, 104–5, 131, 144–45

Krogman, Wilton, 122–23

Kroodsma, Donald, 140–41, 163

Kunst, Jaap, 98

kymograph, 93–97, *94, 96*, 100–101, 103, 105, 126, 136–38, 142

laboratories: *Apparate*, 93–95, 103, 138, *139*; for birdsong study, 127–34, 157–58; and ethnomusicology, 141–42; and experiments, 89, 116, 126, 177; for independent research, 65–73; and institutional sponsorship, 73–75; and intelligent reflection, 85–88, 102–3; musical, 88–92; overview, 4–6;

psychological, 31–32, 85, 88–92; and song typologies, 55. *See also* vivisection

Lach, Robert, 17, 29

"La donna è mobile" (Verdi), 1–2, *2*, 10

language, 22–26, 30–31, 113, 189n86

Laura's Woodland Warbler *(Phylloscopus laurae)*, 78, *79*

Lee, Stan, 162

"Le merle de roche" (Messiaen), 159, *160*

*Le Règne Animal* (Cuvier), 53

Lévi-Strauss, Claude, 155–56

*L'Évolution créatrice* (Bergson), 23

*Life at the Zoo* (Cornish), 20–21, 28, 185n14

*The Life of Reason* (Santayana), 10

List, George, 135, 141

listening, 20–21, 32–34, 35–38, 125–26, 129–32, 135–38, 140–41, 146–48, 164, 205n41

living animals, 46, 57, 68, 85–86, 89, 95, 138

Lockwood, Samuel, 20–21

Loeb, Jacques, 130

Lord Kames, 30

Marine Biological Laboratory, Woods Hole, 65–67

Marler, Peter, 132, 140–41

Marson, Charles, 54

Mason, Daniel Gregory, 29

Mathews, Ferdinand Schuyler, 1–2, 33–34, 49, 168

Mayr, Ernst, 72, 119, 162–63

Mead, Margaret, 119

mechanical transcription devices, 126–42

melody: and aesthetics, 31; and birdsong, 22; and evolution, 19–20, 26; and Gestalt psychology, 137, 216n52; and graphic representation, 100–101; and listening, 168; and mechanical transcription, 128, 133, 140–41; and scores, 49; and song types, 54–55; and the wood pewee, 33–35, 72

melogram excerpt, *134*

melographs, 133–35, 138, 140–42

Messiaen, Olivier, 146–47, 154–62, 166–67

*Mikrokosmos* (Bartók), 57

phrases (musical), 33–34, 63–64, 70–72, 98, 129–30, 140, 164

*Phylloscopus laurae* (Laura's Woodland Warbler), 78, *79*

physiology, 85, 89, 91–92, 93, 97–99, 103, 130–31, 137–39

pigeon studies, 64, 66–69

Pinker, Steven, 25

Pluciennik, Mark, 187–88n54

posthumanism, 7, 9, 12–13, 169–74

postmodern birdsong, 164

postmodern humanity, 109–11, 123–24, 147–48, 152, 165, 166–67, 176–78

postwar Western culture, 110–11

Potter, Ralph, 126, 133

*The Power of Sound* (Gurney), 26

prescriptive notation, 135, 215n35

primitive song, 18, 27–30, 32, 38, 55, 100

*The Principles of Sociology* (Spencer), 18

print discourse, 19–22

Pulitzer, Ralph, 75–77

Pulitzer Angola Expedition, 75–80

race: and culture, 120; and difference, 61–63, 110–11, 115, 121–22; and genetics, 5, 115–16; and identity, 120; postwar, 110–14, 123–24; and racial science, 18, 119–22; and social evolutionism, 27–30; and sonic specimens, 42–43; and typologies, 54–55, 110, 114, 119

racial evolutionism, 119–23, 165, 177

racism, 105, 113, 119–20, 123, 143, 169, 177–78

Radick, Gregory, 23, 32, 127–28

*Rainforest Soundwalks* (Feld), 146–47, 154, 159–62

*The Raw and the Cooked* (Lévi-Strauss), 155–56

Reardon, Jenny, 111

recorded guides, 150–52

"The Rhythmical Song of the Wood Pewee" (Oldys), *35*

Riddle, Oscar, 109–10

ring dove *(Turtur risorius)*, 67–68, 198–99n18

Roberts, Helen, 54–56

robin, American *(Planesticus migratorius)*, 38

Sachs, Curt, 49, 114, 116, 119

Santayana, George, 10, 13

Saunders, Aretas, 1, 34–35, 37–38, 54, 72

Schafer, Murray, 147, 154–55, 161

Schuchert, Charles, 53

Schull, George, 117

*Science* (journal), 64–65, 70

Scott, Édouard-Léon, 95

Scott, Joan, 173

Seashore, Carl, 54–55, 74, 100, 137–38

Seeger, Charles, 125–27, 132–35, 138, 140–41, 143

sexual selection, 24–26, 29, 30, 32, 37

seyak, 154, 161–62, 166, 223n71

Sharp, Cecil, 54

Smocovitis, Vasiliki, 116, 118–19, 211n57, 211n59

social evolution, 27–30, 35–38, 188n54

song collection(s): and Boulton, 74–80; and Craig, 67–68, 70, 72–73; and difference, 61–63, 80–83, 166–67; in the field, 56–58; and guides, 149–52; history of, 62, 113; language of, 113; and museum recordings, 42, 46–50; and notation, 49–50; and typologies, 54–56; and vivisection, 84, 86–87, 91–92, 97–100; and wax cylinders, 92; Westernized, 49–50

*The Song of the Wood Pewee* (Wallace), 72–73, 144

song sparrow, 1–2, 10, 54, 224n1

sonic bodies, 48–50, 54

sonic difference, 61, 80

sonic identity, 152, 158–65

sonic landscapes, 152–56

sonic specimens: audiotyping, 50–56; vs. biological specimens, 45–46, 50–51, 54; in the field, 56–58; institutionalized, 42–43, 46–50; overview, 42–44; and visual information, 46

sonic taxonomies, 81–83

sonic typologies, 42, 131

sonograph. *See* sound spectrograph

Sonoran Desert pop-up scene, *165*

*Sound and Sentiment* (Feld), 161–62

sound recordings, 42, 46–50, 75–76, 80, 81, 92–97, 100–101, 148–49

soundscapes, 76, 146–47, 153–55, 156–65, 166–67

sound spectrograph, 126, 127–33, 138, *139*, 139–41, 151, 217n65

sound truck, Cornell, 54, 150–51

species: classification, 49, 51–53; and culture, 177–78; defined, 44; discovery of, 78; identification, 45, 149–51, 164, 166–67; identity, 45, 68, 118; source recordings, 164; taxonomies, 161–62

specimen collections, 41–43, 46–59, 68, 73–80, 148–52

specimen typing, 150

spectrograph. *See* sound spectrograph

Spencer, Herbert, 17–19, 22–32, 37, 155

spiritual language of music, 22, 36–37, 186n24

St. Vincent Millay, Edna, 114–15

Steward, Julian, 211n67

Strasser, Bruno, 75

Stratton-Porter, Gene, 57–58

Stumpf, Carl, 31, 88–91, 97–98, 203–4n22

"The Subjective Space System" (Craig), 144

subjectivity, 67, 103–4, 127, 130–32, 135–38, 141–43, 144–45, 171, 178–79

Sully, James, 20, 27, 30, 32, 36

"Swanee River" (Foster), 33–35, *35*

synthesis movement, evolutionary, 110, 118–22, 207n8

taxidermy, 195n52

taxonomies: of difference, 73, 174–75, 179; of identity, 159, 173–75; and listening, 164; of music, 61; and naming of species, 161–62; natural, 52; sonic, 81–83; visual, 52, 151; Western, 162

"Thai oi Kamen," 84, 92, 97–100, 104

Thai song vivisection, 84, 92, 97–100, 104

*The Music Hunter* (Boulton), 77

Thone, Frank, 53

Thorpe, William Homan, 125–32

topographic map, 70, *70*, 155

*Torture Chambers of Science (Die Folterkammern der Wissenschaft)* (von Weber), 88

transcription: ethnographic, 50; and the kymograph, 93; mechanical devices for, 126–42; and Messiaen's *Catalogue d'oiseaux*, 156–58; of mouse music, 20–21, 24; and musical type, 54; and vivisection of song, 97–100; on wax cylinders, 46–49; and wood pewee study, 64, 67–70

Tuan, Yi-Fu, 155

type specimens, 45–46, 50–56, 78, *79*, 118–19

typologies: field vs. institution, 56; and handmade graphs, 136; history of, 51–53; prewar, 167; racial, 54–55, 110, 114, 119; and song collection, 54–56; sonic, 42, 131; and type specimens, 46; visual, 149

UCLA (University of California, Los Angeles), 125, 134

unequal relationships, 7, 73–83, *82*

UNESCO Statement on Race, 119–21

University of Berlin, 88–89, 105

University of Chicago, 64–68

University of Iowa, 137–38

University of Maine, 64, 66–68

unpicturables, 97–101, 103

valuation, of animal life, 3–7, 10, 103, 176

Verdi, 1–2, *2*, 10

Virchow, Rudolf, 88

visual guides, 148–52

visual information, 42–43, 46, 48–50, 52, 92–103, 126–29, 131–33, 135–42, 164

vivisection, 84–92, *94*, 95–101, 102–5, 175–76

vocalizations: and behavior, 66–68, 71–72, 140, 189n88; vs. ethnomusicology, 142; and identification, 68; and language, 23–26, 30–31; and social evolutionism, 30, 37; and the spectrograph, 132, 139–40

Vyn, Garret, 163–64

Wallace, Alfred Russel, 25

Wallace, William, 36

Wallaschek, Richard, 31, 36

Washburn, Sherwood, 118

Watson, John B., 130

wax cylinders. *See* cylinders

Weber, Ernst von, 88

"Weii" cylinder study, *99*

Weismann, August, 25, 27–28, 36

Wertheimer, Max, 90, 131

Westernized song collections, 49–50

Western musical notation, 20, 55, 98, 128

Western taxonomies, 162

"Wheel Song," 20, *21*

whisper song, 38, 191n119

whistling, 217n67

Whitman, Charles Otis, 65–68

Witchell, Charles, 25, 128–29

women, in science, 63, 68–69, 73–74

*Wood Notes Wild: Notations of Bird Music* (Cheney), 149

wood pewee, 33–35, 64, 69–73, 81, 140, 144

Wundt, Wilhelm, 89, 91

Yerkes, Robert, 69, 73

Marié Abe
*Resonances of Chindon-ya:
Sounding Space and Sociality in
Contemporary Japan*

Frances Aparicio
*Listening to Salsa: Gender, Latin Popular
Music, and Puerto Rican Cultures*

Paul Austerlitz
*Jazz Consciousness: Music, Race,
and Humanity*

Harris M. Berger
*Metal, Rock, and Jazz: Perception and the
Phenomenology of Musical Experience*

Harris M. Berger
*Stance: Ideas about Emotion, Style,
and Meaning for the Study
of Expressive Culture*

Harris M. Berger and
Giovanna P. Del Negro
*Identity and Everyday Life: Essays
in the Study of Folklore, Music,
and Popular Culture*

Franya J. Berkman
*Monument Eternal: The Music
of Alice Coltrane*

Dick Blau, Angeliki Vellou Keil,
and Charles Keil
*Bright Balkan Morning: Romani Lives and
the Power of Music in Greek Macedonia*

Susan Boynton and Roe-Min Kok, editors
*Musical Childhoods and the Cultures
of Youth*

James Buhler, Caryl Flinn,
and David Neumeyer, editors
*Music and Cinema*

Thomas Burkhalter, Kay Dickinson,
and Benjamin J. Harbert, editors
*The Arab Avant-Garde: Music, Politics,
Modernity*

Patrick Burkart
*Music and Cyberliberties*

Julia Byl
*Antiphonal Histories: Resonant Pasts in
the Toba Batak Musical Present*

Raymond Knapp
*Symphonic Metamorphoses:*
*Subjectivity and Alienation in Mahler's*
*Re-Cycled Songs*

Laura Lohman
*Umm Kulthūm: Artistic Agency and the*
*Shaping of an Arab Legend, 1967–2007*

Preston Love
*A Thousand Honey Creeks Later:*
*My Life in Music from Basie to Motown—*
*and Beyond*

René T. A. Lysloff and
Leslie C. Gay Jr., editors
*Music and Technoculture*

Allan Marett
*Songs, Dreamings, and Ghosts:*
*The Wangga of North Australia*

Ian Maxwell
*Phat Beats, Dope Rhymes: Hip Hop Down*
*Under Comin' Upper*

Kristin A. McGee
*Some Liked It Hot: Jazz Women*
*in Film and Television, 1928–1959*

Rebecca S. Miller
*Carriacou String Band Serenade:*
*Performing Identity in the*
*Eastern Caribbean*

Tony Mitchell, editor
*Global Noise: Rap and Hip-Hop*
*outside the USA*

Christopher Moore and
Philip Purvis, editors
*Music & Camp*

Rachel Mundy
*Animal Musicalities: Birds, Beasts,*
*and Evolutionary Listening*

Keith Negus
*Popular Music in Theory: An Introduction*

Johnny Otis
*Upside Your Head! Rhythm and*
*Blues on Central Avenue*

Kip Pegley
*Coming to You Wherever You Are:*
*MuchMusic, MTV, and Youth Identities*

Jonathan Pieslak
*Radicalism and Music: An Introduction*
*to the Music Cultures of al-Qa'ida, Racist*
*Skinheads, Christian-Affiliated Radicals,*
*and Eco-Animal Rights Militants*

Matthew Rahaim
*Musicking Bodies: Gesture and Voice*
*in Hindustani Music*

John Richardson
*Singing Archaeology:*
*Philip Glass's "Akhnaten"*

Tricia Rose
*Black Noise: Rap Music and Black Culture*
*in Contemporary America*

David Rothenberg and
Marta Ulvaeus, editors
*The Book of Music and Nature: An*
*Anthology of Sounds, Words, Thoughts*

Nichole Rustin-Paschal
*The Kind of Man I Am: Jazzmasculinity*
*and the World of Charles Mingus Jr.*

Marta Elena Savigliano
*Angora Matta: Fatal Acts*
*of North-South Translation*

## ABOUT THE AUTHOR

Rachel Mundy is associate professor of music in the arts, culture, and media department at Rutgers University in Newark.